Nuclear Power Reactor Development

Yoshiaki Oka

Nuclear Power Reactor Development

History, Technologies, and Lessons

Springer

Yoshiaki Oka
Graduate School of Engineering
Emeritus Professor, The University
of Tokyo
Bunkyo City, Tokyo, Japan

ISBN 978-981-96-2730-1 ISBN 978-981-96-2731-8 (eBook)
https://doi.org/10.1007/978-981-96-2731-8

Translation from the Japanese language edition: "Hatsudenyo Genshiro No Kaihatsu: Rekishi, Gijyutu, Kyokun" by Yoshiaki Oka, © Yoshiaki Oka 2024. Published by Amazon. All Rights Reserved.

The original submitted manuscript has been translated into English. The translation was done using artificial intelligence. A subsequent revision was performed by the author(s) to further refine the work and to ensure that the translation is appropriate concerning content and scientific correctness. It may, however, read stylistically different from a conventional translation.

This Springer imprint is published by the registered company Springer Nature Singapore Pte Ltd.
The registered company address is: 152 Beach Road, #21-01/04 Gateway East, Singapore 189721, Singapore

Preface

- How was the development of early power reactors in the United States conducted?
- Why did light water reactors dominate the market for power reactors worldwide?
- How has light water reactor technology progressed?

Many of the new types of reactors currently being discussed have been considered in the past. The history of reactor development is unraveled based on reactor theory, and lessons for the future are considered.

- What are the characteristics of light water reactors in various countries?
- What is the history of the development and use of heavy water-moderated reactors and graphite-moderated reactors?
- Why is the power coefficient likely to become positive in graphite-moderated and heavy water-moderated pressure tube reactors with light water cooling?
- Will uranium resources be depleted?
- What is the history of fast reactor development?
- What are the political aspects of reprocessing and plutonium use?
- What is the history of liquid fuel reactors, molten salt reactors, and organic material-cooled reactors?
- What is the development and history of nuclear reactors for aerospace use and isotope power sources for space use?
- What are third generation (Generation III) and fourth generation (Generation IV) nuclear reactors?
- What is a supercritical pressure light water reactor (SCWR)?
- What is the overall picture of the conceptual design and analysis methods for power reactors?
- What are fusion reactors and fission-fusion hybrid reactors?
- What are the differences between nuclear engineering education in American and Japanese universities?
- What is the knowledge base for nuclear power utilization and research and development?

The United States actually built and examined various reactors in the 1950s and 1960s. Many of the reactors currently being researched and developed around the world have their roots in these. This book introduces not only the power reactors that were actually built, but also those that were conceptually considered.

The history of research and development of power reactors worldwide is comprehensively summarized, from light water reactors to fusion reactors, based on author's own research and experiences gained through contact with development in Japan and around the world.

This book not only presents the history of power-generating nuclear reactors based on research findings, but also discusses the results and lessons learned from logical considerations based on nuclear reactor theory as a nuclear reactor design expert. For example, it discusses questions such as "Why is the coolant void coefficient likely to be positive in a pressure tube type nuclear reactor with light water cooling, graphite moderation, or heavy water moderation?", "Why did Canada abandon operation of the MAPLE-X10 built for isotope production when it was found during pre-use inspection that the power reactivity coefficient was slightly positive during startup?", "Why is the power density of graphite-moderated reactors and heavy water-moderated reactors lower than that of light water reactors?", and "Why is the power density of liquid metal-cooled fast reactors larger than that of light water reactors, but the construction cost higher?".

Many nuclear power experts are specialists in one of the fields such as nuclear reactor physics, heat transfer, fluid engineering, materials science, structural mechanics, chemical engineering, and radiation chemistry, and have little experience in considering and acting across fields. In this book, the author introduces in some detail the overall picture of the design and analysis of power-generating nuclear reactors, using the supercritical pressure light water reactor that he has researched for 25 years as an example. It should serve as a reference for the design and analysis methods of reactor cores, plant control, stability, safety systems, etc. of light water reactors and the like.

In addition to technology, the book also discusses investment risks in power-generating nuclear reactors and ways to reduce electricity costs. It also introduces nuclear engineering education in Western universities, as the differences with Japan are not well known. It also discusses the knowledge base for nuclear power utilization.

While this book points out issues and problems with references to the literature, it does not suggest "what should be done". This is because those who have the budget, authority, or responsibility for use should think and act. If someone who does not fully understand the whole situation easily says "this should be done", it can cause failure and make responsibility unclear. As a measure to make the policy and execution results of the responsible agency (administrative agency) robust, it also points out that there is a mechanism in the West to question the policy and execution results (accountability) of the responsible ministries and organizations, which is not unique to the nuclear power field, but is lacking in Japan.

The book hopes to deepen understanding of power-generating nuclear reactors, prevent the same mistakes, think about solutions, and serve as a reference for future research and development.

The intended readers of this book are those who are interested in nuclear power generation. For those already involved in nuclear power generation, it should be useful in thinking about the operation and improvement points of use and research and development. It is also expected to be useful in passing on experience to the next generation by using it as a textbook, exercise book, or supplementary reader at university.

Furthermore, issues at the intersection of nuclear power generation and society, such as economics, safety and communication, environment, and nuclear law systems, are described in the companion book "Nuclear Power and Society, Issues on Economy, Safety, Environment and Law". The author hopes readers will refer to it along with this book. The issue of accountability is discussed in Chaps. 4 and 5 of the book.

Bunkyo City, Japan Yoshiaki Oka

Competing Interests The author has no competing interests to declare that are relevant to the content of this manuscript.

Introduction

This book investigates, compiles, and discusses the history and lessons of the research and development of power reactors worldwide from the dawn of nuclear energy use to the present, based on reports from Western government agencies and others.

The development of reactors for the use of nuclear energy for power generation and propulsion was actively carried out in the United States after World War II. The military, the Department of Energy, and companies built various test reactors and power reactors. Not only light water reactors, but also what are now called new types of reactors, most of them have their roots in these. The Pressurized Water Reactor (PWR) was developed based on a submarine reactor, and the Boiling Water Reactor (BWR) was developed by an American electric company in search of a type of light water reactor different from the pressurized water reactor. For testing purposes, the U.S. Department of Energy built a nuclear reactor test site in the desert of Idaho. Various types of reactors are being tested, not only for power generation but also for aircraft and space propulsion, marine use, and remote power sources. In the UK, France, Canada, and the Soviet Union, unique power-generating nuclear reactors were developed.

Understanding the history and lessons of research and development of power-generating nuclear reactors is as important as studying nuclear reactor physics and heat transfer fluid engineering. However, while there are books that describe the structure, dimensions, and design standards of light water reactors, it is rare to find ones that describe the history and circumstances leading up to them, not as a story, but as a record. The experiences of accidents and troubles with early light water reactor test reactors and prototype reactors are reflected in the design of the light water reactors currently in use. Those who started using light water reactors in Japan had heard about the experiences and design changes in the early stages of light water reactor development when they built and operated the first power-generating units of the Power Reactor (JPDR), PWR, and BWR, and they had that knowledge. However, they are no longer with us, so we cannot ask them about their experiences. Not only the reactors themselves, but also the technologies used in power-generating nuclear reactors have been improved and advanced. These include nuclear fuel, instrumentation and control technology, and nuclear reactor analysis technology. Improvements to enhance passive safety to reduce dependence on power-requiring active equipment, and measures against core meltdown accidents, have been researched and developed worldwide since the Chernobyl accident. This book will also introduce these technologies and developments.

Currently, light water reactors have become the mainstream of power-generating nuclear reactors, but various types of power-generating nuclear reactors were developed and used worldwide. These include the Heavy Water-Moderated Reactor (CANDU) developed by Canada, the Graphite-Moderated Carbon Dioxide-Cooled Reactor developed by the UK and France, the High-Temperature Gas Reactor of the United States and West Germany, the Graphite-Moderated Boiling Light Water-Cooled Pressure Tube Reactor (RBMK) of the former Soviet Union, and the Liquid Metal-Cooled Fast Reactor developed in many countries. Some of these are still in use and being researched and developed. Furthermore, although they have not yet reached full practical use, liquid fuel reactors, molten salt reactors, organic material-cooled reactors, and nuclear reactors for aerospace use have also been researched and tested. Among these, there are similar types that are being discussed in research and development as new types of reactors. Knowing the history of these reactor types, and the lessons and challenges of their development, is necessary to broaden the perspective on power-generating nuclear reactors, recognize the challenges, and plan for the future.

Around the year 2000, under the leadership of the United States, practical applications were made toward the construction of generation three (Generation III) nuclear reactors, and research and development programs for generation four (Generation IV) nuclear reactors were conducted. The supercritical pressure light water reactor (SCWR) that the author researched at a university from the late 1980s was selected as a Generation IV nuclear reactor and was researched worldwide. The author introduced this experience in this book because he thought it would be a reference for the future. Since being appointed as the Chairman of Japan Atomic Energy Commission of Japanese government in 2014, the author has shifted his focus and interest to the issues at the intersection of nuclear power generation and society, and has not conducted research on SCWR.

At the end of this book, after briefly introducing the history of conceptual research on fusion reactors and fusion-fission hybrid reactors, the author discusses the knowledge base for nuclear power engineering education at universities and the use of nuclear power, which are related to research and development. Japan was the first in the world to develop computer-aided design, construction, and maintenance technology for light water reactors, but this is introduced in another book, "Nuclear Power and Society", so it is not mentioned in this book. In "Nuclear Power and Society", the author investigated and described the history and current situation of nuclear power use, such as economic efficiency and investment risk, stable power supply, safety and communication, radioactive waste and environment, nuclear safety regulation, disaster prevention, damage compensation, and non-proliferation, based on information from government agencies and others. It would be appreciated if you could refer to the book along with the present book. The author also briefly mentions the reduction of investment risk and power generation cost in Chap. 15 of this book, as they are related to the research and development of power reactors.

At the end of the volume, exercise problems are provided. It would be appreciated if you could use them to help recognize and understand the issues. The information circulating in the public, both against and in favor of nuclear power, is often partial and biased. Instead of taking these at face value, the author recommends searching in English, verifying the basis, and thinking for yourself. This book is not only for university and graduate students studying nuclear engineering, but also for those involved in the use of nuclear power generation, such as power companies, nuclear reactor manufacturing and construction companies, research and development institutions, university faculty, administrative officials of ministries and local governments, and the general public who are broadly interested in nuclear power generation. The author hopes that this book will serve as a reference for the future use and research and development of power reactors.

Contents

About the Author

Yoshiaki Oka is Professor Emeritus at the University of Tokyo, and former chairman of the Japan Atomic Energy Commission (2014–2020). His specialty is nuclear engineering, and he has studied the concepts and analysis methods of various nuclear reactors based on knowledge of nuclear reactor physics, radiation transport, and nuclear reactor thermal fluid dynamics. Since taking charge of the Japan Atomic Energy Commission, he has been examining issues at the intersection of nuclear power generation and society.

He graduated from the Department of Nuclear Engineering, Faculty of Engineering, University of Tokyo in 1969, completed master's and doctoral programs of the University of Tokyo. In 1974, and became an assistant professor of Nuclear Engineering Research Laboratory of the Faculty of Engineering of the University of Tokyo, then an associate professor, and in 1989, a professor. He was engaged in research and education in nuclear engineering at the University of Tokyo for 36 years. He was also responsible for the operation and joint use of the University of Tokyo's nuclear research reactor facilities. His specialty is nuclear reactor design engineering, and he has studied various nuclear reactor concepts. The supercritical pressure light water reactor he developed was selected as a Generation IV nuclear reactor in 2000 and is being studied worldwide. In 2005, he led the establishment of the Nuclear Engineering and Management Major and the Nuclear Professional School based on the faculty quota of the Nuclear Engineering Research Laboratory. Based on the education, he led the creation of the Nuclear Engineering Textbook Series (Ohmsha in Japanese, Springer in English), and also authored and edited it by himself. In 2007, he was the representative of the Ministry of Education, Culture, Sports, Science and Technology's Global COE Program in the field of nuclear power. In 2010, he retired from the University of Tokyo and was engaged in the establishment and operation of the Cooperative Major in Nuclear Energy of Waseda University. From 2014 to 2020, he served as the chairman of the Japan Atomic Energy Commission, which was reorganized after the TEPCO accident. He served as the President of the Atomic Energy Society of Japan in 2008 and a member of Board of Directors of the American Nuclear Society between 2001 and 2004.

He authored and edited several books in English and Japanese such as "Nuclear Reactor Design" (Ohmsha, 2010, Springer 2014) and "Super Light Water Reactors and Super Fast Reactors" (Springer, 2010). He has published about 230 research papers in English. He is ranked as the second scholar in the field of nuclear reactors and the 16th scholar in the field of nuclear power on ScholarGPS™.

Abbreviations

ABWR	Advanced Boiling Water Reactor
ANL	Argonne National Laboratory
BWR	Boiling Water Reactor
CANDU	Canada Deuterium Uranium Reactor
CEA	French Alternative Energies and Atomic Energy Commission
EDF	Électricité de France
EPR	European Pressurized Water Reactor
EPRI	Electric Power Research Institute
GAO	Government Accountability Office
GE	General Electric Companies
GIF	Generation IV International Forum
IAEA	International Atomic Energy Agency
INEEL	Idaho National Engineering and Environmental Laboratory
INL	Idaho National Laboratory
MOX	Mixed Oxide Fuel
NRC	Nuclear Regulatory Commission
OECD/NEA	Organization for Economic Cooperation and Development/Nuclear Energy Agency
ORNL	Oak Ridge National Laboratory
PWR	Pressurized Water Reactor
RBMK	Reaktor Bolshoy Moshchnosti Kanalnyy
SCWR	Supercritical Water Reactor
VVER	Vodo-Vodyanoi Energetichesky Reactor
WH	Westinghouse Electric Corporation
WNA	World Nuclear Association

Chapter 1
Early Reactor Development in the United States

This chapter briefly introduces the early development of nuclear reactors in the United States, and then describes the development and commercialization of pressurized water reactors and boiling water reactors in the United States.

In the United States, during World War II, a nuclear reactor was built and operated in Hanford, Washington, using graphite as a neutron moderator to produce plutonium for nuclear weapons. Subsequently, power reactors using graphite as a moderator were built in the UK, France, the former Soviet Union, the United States, and China.

The U.S. Atomic Energy Commission assigned the Argonne National Laboratory the role of developing commercial nuclear power reactors from the 1940s. The Argonne Laboratory built numerous experimental reactors on the grounds of its suburban Chicago laboratory and at the Atomic Reactor Test Site in Idaho, known as the Argonne West Laboratory (renamed the Idaho National Laboratory from 2005). In Idaho, the Experimental Breeder Reactor-I (EBR-I) generated the first electricity from nuclear energy in 1951, lighting four light bulbs. EBR-I is a fast reactor that uses plutonium as fuel and was the first to demonstrate the ability to produce more fuel than it consumes. The first to send nuclear power electricity to a nearby town was an experimental boiling water light water reactor called BORAX-III, also built in Idaho, in 1955 (Touran 2020; Simpson 1994).

The development of nuclear reactors for nuclear submarines was carried out by Admiral Rickover of the Navy. Initially, a reactor cooled by liquid sodium and using beryllium as a moderator was tested, but it did not work well, and a pressurized water light water reactor was developed. The full-scale test reactor, Submarine Test Reactor (STR) became critical in 1953, and in 1955 the first nuclear submarine Nautilus was launched, proving the advantages of nuclear power that can navigate for a long time without refueling. The liquid sodium-cooled test reactor SIR was built by General Electric for the nuclear submarine Seawolf, but it did not show satisfactory performance, and, since then, nuclear reactors for nuclear submarines have been pressurized water light water reactors. Those with increased output are used on nuclear aircraft carriers. For civilian use, the nuclear-powered merchant

Y. Oka, *Nuclear Power Reactor Development*,
https://doi.org/10.1007/978-981-96-2731-8_1

ship Savannah was built and sailed for about 10 years from 1962. Civilian nuclear ships have not been practical due to the small cargo space and the need for many crew members. All of these reactor types are pressurized water light water reactors. It should be noted that the output of shipboard power reactors is much smaller than that of power generation reactors. Among shipboard reactors, those for aircraft carriers have a larger output than those for submarines.

In the United States, development of nuclear engines for cruise missiles and supersonic transport aircraft was also carried out. Development was carried out for about 10 years from around 1950, but due to the advancement of intercontinental ballistic missile (ICBM) technology, development was discontinued in the early 1960s (Stacy 2000, Chap. 13). The molten salt reactor, in which the liquid (molten salt) nuclear reactor fuel also serves as a coolant, originated from this development.

The development of nuclear reactors for space use was carried out in the United States and the former Soviet Union in the 1960s. In deep space exploration where sunlight does not reach, nuclear power is necessary, and low-output ones use radioisotopes such as plutonium as a heat source, and high-output ones use nuclear reactors as a heat source, and electricity is mainly obtained by thermoelectric conversion. It is based on the Seebeck effect where electromotive force occurs in metals and semiconductors with temperature difference. The U.S. program is called the SNAP (DOE 2015).

The U.S. Army developed an ultra-small nuclear reactor in the 1950s for use in remote locations such as polar bases. It was assumed that it would be manufactured in a factory and used after being transported to the polar regions. The type of nuclear reactor is mainly a pressurized water light water reactor. Remote power sources using fossil fuels often require fuel transportation, but nuclear power does not.

In 1953, President Eisenhower declared "atomic energy for peace", and the civilian use of nuclear energy began. The development of civilian nuclear reactors in the United States began in 1954 when the U.S. Atomic Energy Commission made a five-year plan to create and examine the following five experimental nuclear reactors: Shippingport PWR, Experimental Boiling Water Reactor (EBWR), Sodium Reactor Experiment (SRE), Homogeneous Reactor Experiment (HRE-2), and Experimental Breeder Reactor (EBR-II). Of these, Shippingport PWR and EBWR were later put into practical use as pressurized light water reactors and boiling light water reactors. SRE and EBR-II are liquid sodium-cooled reactors, and EBR-II is a fast reactor using metal fuel that has been operating at the Idaho National Laboratory for 30 years from 1964. The reprocessing of spent metal fuel from EBR-II was done by high-temperature metallurgy. This reprocessing method is also called electrolytic refining because it melts and refines the fuel by electrolysis. It is also called a dry method, as opposed to the wet method of dissolving spent fuel in nitric acid solution.

In the United States, the Atomic Energy Act of 1954 made it possible for private companies to own nuclear reactors. The Atomic Energy Commission implemented the Nuclear Power Demonstration Program from 1955 to 1960, and various types of nuclear reactors were built and operated by consortia of private companies such as power companies. These included liquid sodium-cooled graphite-moderated reactors, organic material-moderated and cooled reactors, heavy water-moderated pressure tube reactors, gas-cooled heavy water-moderated pressure tube reactors, and nuclear superheated boiling water light water reactors, all of which were shut down within a few years. The liquid sodium-cooled fast reactor (Fermi 1) and the helium-cooled graphite-moderated reactor (Peach Bottom) operated for about 10 years from the initial criticality but were shut down in the 1970s. In contrast, the light water reactors built at that time, both pressurized water and boiling water types, were in operation until the 1980s or 1990s. These include the Shippingport PWR, Yankee Rowe PWR, Big Rock Point BWR, La Crosse BWR, and San Onofre PWR (Allen 1977).

Improvements in the design of light water reactors up until the 1980s are detailed in L. S. Tong's book (Tong 1988). For technology related to nuclear power generation, there is, for example, a guidebook compiled by authors from the Electric Power Research Institute (Rahn et al. 1984).

1.1 Development and Commercialization of Pressurized Water Reactors

The Pressurized Water Reactor (PWR) was developed by the U.S. Navy for use in nuclear submarines and was later upscaled for practical use. PWRs are the most commonly used for nuclear power generation worldwide.

The Shippingport PWR was an engineering test reactor for power generation built in the suburbs of Pittsburgh, Pennsylvania. It became critical in 1957 and was operated until 1982. The electrical output was 60,000 kilowatts. The first core was a seed-blanket core, which used a highly enriched uranium seed part surrounded by a blanket part made of natural uranium, repurposed from a canceled nuclear aircraft carrier project. The final, third core was a thermal neutron breeding reactor, with the seed part made of Uranium-233 and the blanket composed of thorium. The construction cost per power output was about ten times higher than that of oil or coal fired power plants, so the Shippingport PWR is referred to as an engineering test reactor, not a commercial prototype.

The Yankee Rowe PWR, with a power output of 185,000 kilowatts (initially 110,000 kilowatts), was the first commercial pressurized water reactor funded and built by ten power companies in New England. It began operation in 1960 and ran until 1992 with an exceptionally high average operating rate of 74% for a prototype. It was the first to use uranium oxide fuel, which is still used in light water reactors today. The cladding was made of stainless steel, as zirconium alloy cladding was not yet available. A spherical steel containment vessel was used. The Yankee Rowe PWR was completed at 23% less than the initial estimated construction cost, making the commercial use of nuclear power a reality. Figure 1.1 shows a photo of the Yankee Rowe power plant.

Fig. 1.1 Yankee Rowe nuclear power plant. *Source* Nuclear regulatory commission from US–Yankee Rowe

Subsequently, the Indian Point PWR (163,000 kilowatts) started operation in 1962, the San Onofre PWR (436,000 kilowatts) in 1967, and the Connecticut Yankee PWR (560,000 kilowatts) in 1968. They were operated until 1974, 1992, and 1996, respectively. Note that larger second and third units were later built at Indian Point and San Onofre.

The success of these initial units led to a construction rush of nuclear power plants in the United States, as if "the dam had been broken". This boom in the construction of nuclear power plants in the United States continued until the 1970s, but many companies suffered losses due to delays in construction periods at power companies lacking project management capabilities. As a result, new construction in the United States has significantly slowed down since the 1980s. The Three Mile Island nuclear accident in 1979 also had an impact. Furthermore, the deregulation of electricity in the United States from the late 1980s added to the difficulties of new construction. The completed light water reactors have been providing cheap electricity since the 1990s, after improving their operating rates. The U.S. government has promoted regulatory improvements and, in the 2000s, has backed new construction by guaranteeing part of the construction costs, leading to the construction of Vogtle Units 3 and 4.

In the research and development of PWRs, Westinghouse, the manufacturer, built the Saxton power test reactor (thermal output 20,000 kilowatts) in 1960 and operated it until 1972, conducting research to improve the economy of PWRs. For example, they developed a chemical volume control (Chemical Shim) method, which involves adding boron, a neutron absorber, to the primary coolant and reducing its concentration along with the burning of nuclear fuel to maintain criticality over a long period. This has reduced the number of control rods and their driving mechanisms in PWRs, which is now used in current PWRs. Irradiation tests of nuclear fuel using a zirconium alloy cladding material called Zircaloy are also being conducted.

The first to adopt the Rod Cluster Control (RCC), which is used in current PWRs, instead of the cross-shaped control rods, was the San Onofre Unit 1. The RCC eliminated the need for control rod followers, which were installed to suppress the peak of the power distribution created when the cross-shaped control rods were pulled out, resulting in a smaller reactor pressure vessel, and the benefits have led to the miniaturization of pressurizers and reactor containment vessels.

The cost of power generation per output of light water reactors decreases as the power output increases. Westinghouse has increased the cooling system of its pressurized water reactors from 2 to 3 or 4, thereby increasing the power output of the pressurized water reactors without changing the design of the steam generator and primary coolant pump. In other words, the number of cooling systems is 2 for a 500 MW class PWR, 3 for a 750 MW class, and 4 for a 1000 MW class, making it a standard plant for PWRs. This allows them to increase the output without changing the design of important equipment such as steam generators and coolant pumps, and to provide PWRs that match the power company's transmission network and demand volume. This standardization is an excellent strategy (Shimada 2008).

WH began developing a floating nuclear power plant in the early 1970s. The motivation is to reduce costs through the standardization of power plants and the need to increase the location of power plants near densely populated areas in the eastern United States. The advantages include improved quality control through factory production, standardization of light water reactor manufacturing, i.e. the design and construction of nuclear power plants are not affected by varying conditions depending on the power plant site, and the constraints on construction sites are reduced. Computer-aided design (CAD) is also being used for the first time. A joint venture with a shipbuilding company has been established. The plan was to place a floating nuclear power plant in the middle of a bank, within 3 miles of the coast of a river mouth or sea. Although orders were received and evaluations were received from the NRC, the U.S. Coast Guard, and the Environmental Protection Agency (EPA), the plan did not materialize due to the impact of the 1973 oil shock, which caused a slump in electricity demand. At the time the decision was made to cancel the plan, permission had been obtained from the NRC to build eight floating plants (Simpson 1994, pp.213–222). Floating nuclear power plants have occasionally been researched and discussed since then.

In the United States, Babcock & Wilcox (B&W), a manufacturer of thermal power boilers, and Combustion Engineering (CE), a manufacturer of large equipment such as nuclear reactor pressure vessels, also started manufacturing pressurized water light water reactors for power generation. The first plants of B&W and CE went into operation in 1973 and 1971, respectively. CE's PWR has two steam generators even for large reactors, while Westinghouse has four. B&W's PWR is a two-loop PWR, and unlike WH's PWR which uses a U-tube type steam generator, it uses a through-flow type steam generator. The Korean type PWR is based on CE's design. The Three Mile Island Unit 1, which had a core meltdown accident in 1979, is a B&W type PWR. B&W also manufactured the nuclear reactor for the U.S. nuclear-powered merchant ship Savannah, which was built in 1961. The related company, German Babcock, partnered with Brown Boveri to supply a nuclear reactor to the German nuclear ship Otto Hahn, and is also building a large PWR in Mülheim-Kärlich, near Bonn.

As of 2017, 79 pressurized water light water reactors have been built in the United States, of which WH has built 55, CE 15, and B&W 9 as the main contractors (EIA 2021). CE was acquired by Westinghouse in 2000. In addition, 39 boiling water light water reactors have been built, all manufactured by General Electric (GE). All of these companies have been involved in the nuclear submarine and nuclear aircraft carrier business from the beginning. The construction of nuclear power plants in the United States and Japan involves not only nuclear reactor manufacturers but also civil engineering and construction companies.

The nuclear division of Westinghouse was acquired by British Nuclear Fuel Limited in 1999, and then purchased by Toshiba in 2006, but was sold to an investment company called Brookfield Business Partners in 2018. WH secured a contract in 2007 to build four PWRs called AP1000s in China, including technology transfer. The construction was delayed, but all four units were operational by 2019. The construction of two AP1000s in the United States began in 2013, but the construction was delayed and WH went bankrupt. In 2017, the power company that placed the order took over the construction from WH, decided on a new construction company, and continued the construction. They are operational in 2023 and 2024, marking the first new nuclear power plants in the United States in a long time.

1.2 Development and Commercialization of Boiling Water Light Water Reactors

The development of boiling water light water reactors began in 1953 when the Argonne National Laboratory built an experimental reactor at the Idaho Atomic Reactor Test Site. A plan called Boiling Water Reactor Experiment (BORAX) sequentially built experimental reactors called BORAX-I to BORAX-V. In BORAX-I, experiments were conducted on the self-control of boiling water light water reactors and the energy generated during nuclear runaway. Initially, it was thought that the steam bubbles generated by boiling might cause output instability, but it was found that the increase in steam bubbles could achieve self-control (inherent safety) that suppresses output increase. After that, about 70 various nuclear runaway experiments were conducted, and the understanding of the energy generated during nuclear runaway was deepened. The final experiment of BORAX-I conducted in 1954 was a destruction experiment of this reactor, which was conducted by rapidly pulling out the reactor control rods. The rapid increase in nuclear reactions granulated the reactor fuel, and a rapid reaction with water caused a steam explosion, destroying the upper part of the reactor. Valuable knowledge about nuclear runaway has been obtained. The photo is shown in Fig. 1.2 (ANL 2023).

Fig. 1.2 BORAX-I nuclear excursion experiment. *Source* Argonne national laboratory

BORAX-II was constructed in 1954, and experiments such as boiling water cooling under pressure and instability research were conducted. When it was built, there was no power turbine, but in 1955 a turbine generator was added, clarifying that radiation contamination on the turbine side is not a major issue. BORAX-III, built in 1955, conducted power generation tests. Equipped with a 20 MW turbine generator, it supplied nuclear power to a suburban town called Arco. This was the world's first event to supply nuclear power to an entire town. BORAX-IV began operation in 1956, conducted tests on uranium–thorium oxide fuel, and clarified that oxide fuel can be used at higher temperatures than metal fuel, that the reaction with water is not as intense as metal fuel even if the fuel cladding tube is damaged, and that it has advantages in manufacturability and high burnup. Uranium oxide fuel is currently used in light water reactors. Tests using fuel rods with holes in the fuel cladding tube were also conducted, and analyses of radioactive material adhesion to turbine side equipment, release of gaseous fission products, radioactive material amount in cooling water, and radiation levels downwind of the reactor were conducted. This has clarified that boiling water type light water reactors can operate safely for a long period of time even if there are defects in many fuel cladding tubes. BORAX-V is studying the stability of high power density cores and nuclear superheating. The former study is conducted in natural circulation and forced circulation of cooling water, and knowledge about the stability of boiling water reactors has been obtained. The latter study was intended to examine the possibility of nuclear power in response to the fact that coal-fired power generation at the time was advancing high-temperature steam temperatures through the use of superheated steam. Cores with a superheater in the center and cores with a superheater in the periphery were studied. In 1964, the latter core was operated for 5 months, and the BORAX-V project was terminated.

GE announced a development strategy named Operation Sunrise in 1958 and embarked on the development of a BWR power plant pursuing economic efficiency. As part of this, the development of a nuclear superheater was carried out until the 1960s. Boiling Nuclear Superheater reactor (BONUS, thermal output 50 MW) was built in Puerto Rico and operated from 1964 to 1968. Pathfinder was built by a power company in South Dakota, but was plagued by superheater trouble, could not demonstrate the expected performance, and was converted to an oil-fired power plant in 1968. Nuclear superheaters did not reach practical use in the United States. These nuclear superheaters operate at subcritical pressure in the reactor, and superheat the steam generated by boiling. It is different from the method of the supercritical water-cooled reactor (SCWR), a Generation IV reactor being considered from the 2000s. The pressure of a supercritical water reactor (SCWR) is different from that of a nuclear superheater. The pressure of the nuclear superheater is subcritical, and boiling phenomena exist, but, in the latter, there is no boiling phenomenon because the pressure of the reactor is supercritical. At supercritical pressure correlation between pressure and boiling temperature disappears, and the cooling water transitions from a high-density state to a low-density state with heating. In other words, there is no need to separate steam from water at supercritical pressure. The nuclear superheater needs to recirculate the high-temperature water after separating the steam. In coal-fired power generation, supercritical pressure power plants have been developed in the United States since the 1950s and are widely used worldwide.

Argonne National Laboratory developed a very small reactor called Stationary Low Power Reactor number 1 (SL-1), with a thermal output of 5 MW, for use in polar regions where it is difficult to transport fossil fuels, and built it at the Idaho National Reactor Testing Station. It succeeded in supplying electricity and heat for heating in 1958. The U.S. Army is participating for performance verification and operator training. SL-1 is known for causing a criticality accident in 1961. The accident occurred when a worker who was performing maintenance during the reactor shutdown lifted a control rod by hand, causing the reactor to immediately become critical and destroyed. Three people, including this worker, died. After this accident, a standard was set for the design of light water reactors, which states that even if a single control rod is pulled out from a stopped reactor, it will not become critical (a state where the nuclear fission chain reaction continues), or in other words, when stopping an operating (critical) reactor, even if a single control rod cannot be inserted due to a malfunction, the reactor can be stopped (made subcritical) by other control rods. It is called one-rod stuck margin criterion, which is used today.

Returning to the development of boiling water light water reactors, following BORAX-IV, the Experimental Boiling Water Reactor (EBWR, electrical output 5 MW) was built at Argonne National Laboratory in the suburbs of Chicago and started operation in 1956 and was operated until 1967. The purpose was to accumulate power tests for long-term power generation and experience with failures and troubles.

The operating pressure of a boiling water light water reactor was 300psi (pound per square inch) in BORAX and SL-1. The pressure was about 20 atmospheres (approximately 300 psi) in the early days, but it has been increased to 600 psi in the

Experimental Boiling Water Reactor (EBWR), and to 1000 psi in commercial power plants up to the present day. The higher the pressure, the higher the temperature of the steam (saturated steam) obtained by boiling, which increases the power generation efficiency. However, it is believed that the sensitivity (the effect of improving power generation efficiency when pressure is increased) is not large, and 1000 psi (6895 kPa) is chosen. The operating pressure of the pressurized water reactor (PWR) is 2250 psi (15513 kPa), and the cooling of the reactor is not done in a saturated boiling state, but in a nuclear boiling state. If the operating pressure is expressed in atmospheres, the boiling water reactor (BWR) is about 68 atmospheres, and the PWR is about 153 atmospheres. The PWR uses high-temperature, high-pressure water heated in the reactor to perform heat exchange in the steam generator, creating saturated steam to rotate the power generation turbine. The pressure and temperature of this saturated steam are about 60 atmospheres and 277 degrees, which is slightly lower than the 68 atmospheres and 286 degrees of the BWR.

General Electric (GE) built a nuclear research institute in Vallecitos, near San Jose, California, and operated the Vallecitos BWR (electric output 5 MW) from 1957 to 1963. This reactor was the first to be built and operated by a private company and to supply power through the civilian power grid. This reactor served as a test reactor for GE to develop BWRs, and tests for improving economy, natural circulation and forced circulation, fuel development and reactor internal structures, control systems, etc. were conducted, and it was used for training operators and workers. Even after the shutdown of this reactor, the Vallecitos Research Institute remained a base for GE's light water reactor research and development for a long time.

The Dresden Nuclear Power Plant Unit 1 (Dresden-1, electric output 180 MW) was the first nuclear power plant to be owned and operated by a civilian power company, and it was operated from 1960 to 1974. Dresden-1 was built in Illinois, and Units 2 and 3 were built on adjacent land in 1970. Its photo is shown in Fig. 1.3. The right side of the photo is Unit 1, the dome on the right side of the exhaust tower is the reactor containment vessel, and the left side is the turbine generator building. In Dresden-1, zirconium alloy (Zircaloy 2) was utilized for fuel cladding tubes. The control rods were improved after experiencing breakage problems, recording a good operational performance with an operating rate of 73% in 1961, demonstrating the practicality of the boiling water reactor. The experience of control rod breakage problems has been used to improve the reactor startup procedure. Dresden-I had a cooling system called a dual cycle, which included a steam drum and a steam generator. The dual cycle is a method that has both a direct cycle (where the steam generated in the reactor is sent directly to the power turbine) and an indirect cycle (where the high-temperature, high-pressure water from the reactor is heat-exchanged in the steam generator to indirectly generate steam). The dual cycle was adopted to improve load following (responsiveness to power demand change). In boiling water light water reactors, like Dresden Unit 1, zirconium alloy fuel cladding tubes have been used from the early commercial power plants, but in the early power plants of pressurized water light water reactors, such as Yankee Rowe and Connecticut Yankee power plants, stainless steel cladding tubes were used throughout the operation period until they were decommissioned.

Fig. 1.3 Dresden nuclear power plants. *Source* NRC

Figure 1.4 shows the transition of the cooling system of the boiling water light water reactor. Dresden-I is a type called BWR-1. BWR-1.5 is a type that removed the steam drum from BWR-1, built at the KRB-A power plant in Gundremmingen, West Germany. BWR-2 is a mainstream type of early boiling water reactors, and it is a single cycle of only the direct cycle where the steam generated in the reactor is sent to the turbine, with the steam generator removed from BWR-1.5. It cools the reactor core with a pump set outside the reactor pressure vessel. Although there are various configurations for BWR-1, since BWR-2 became mainstream, all previous types can be considered as BWR-1. BWR-3, 4, 5 use a jet pump to circulate the core cooling water. With a recirculation pump set outside the reactor pressure vessel, a high-speed water flow is created in the jet pump section inside the reactor pressure vessel, and the suction force drives the feedwater returning from the turbine condenser and the high-temperature, high-pressure water after separating the steam in the steam-water separator cools the reactor. ABWR adopts an internal recirculation pump (internal pump) built into the reactor pressure vessel, and the recirculation system piping at the bottom of the reactor pressure vessel has been removed. This eliminates the need to consider accidents where cooling of the reactor core is lost due to the rupture of pipes below the reactor pressure vessel in safety reviews. The internal pump was adopted in Swedish BWRs in 1971 with the Forsmark Unit 1 having been used. These BWRs were forced circulation types that used pumps to circulate cooling water to cool the reactor core, but in the early days, there were also natural circulation type BWRs that cooled by natural convection without using a pump. In addition to the previously mentioned EBWR and VBWR, the Japan Power Demonstration Reactor (JPDR),

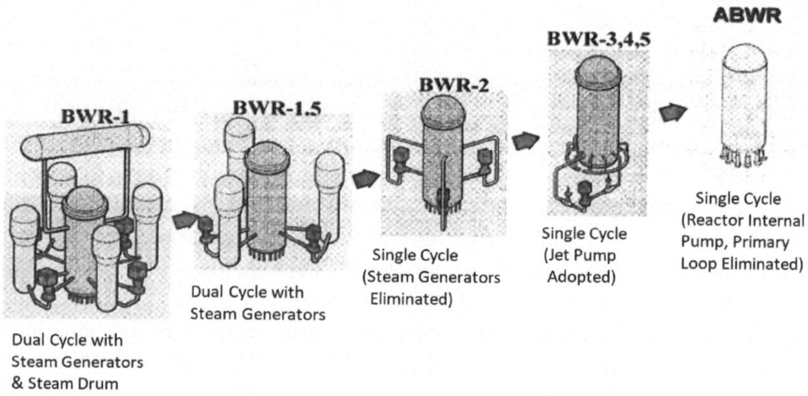

Fig. 1.4 Evolution of boiling water reactors. *Source* Y.Oka (editor) "Advances in light water reactor technologies", Springer 2011

completed in Japan in 1963, was initially a natural circulation type. The JPDR was later operated as a forced circulation type. The Elk River reactor, completed in the United States in 1964, was a natural circulation type with an indirect cycle with a steam generator (Hoshide 2008; Gavrilas 1995).

Dresden Unit 1 was technically successful, but economically it could not compete with the thermal power generation of the time. GE announced the Operation Sunrise plan in 1958 and developed a BWR power plant that pursued economic efficiency. The Big Rock Point Nuclear Power Plant (1963, electric output 75 MW) was the first plant of the BWR-2 type, and it adopted a recirculation flow control that changes the reactor output by changing the cooling flow rate of the reactor core with a recirculation pump, and a method of seeking the optimal pattern of control rods using a computer and optimizing the reactor core power distribution and fuel burnup management is being tested. These methods are used in current BWRs.

At the Humboldt Bay Nuclear Power Plant (1963, electric output 65 MW), the construction cost was reduced by aiming for a simple configuration as much as possible, eliminating pumps and pipes necessary for forced circulation and making it natural circulation cooling, making the steam-water separator a built-in type in the reactor pressure vessel, and adopting a pressure suppression chamber that condenses the steam released during a pipe rupture accident in a water pool, aiming for miniaturization of the containment vessel volume. The built-in steam-water separator in the reactor pressure vessel and the containment vessel with a pressure suppression chamber are used in subsequent BWRs. Natural circulation BWRs did not become the mainstream of subsequent BWRs, but have been attracting attention again in recent years. After the Chernobyl accident, methods with safety that do not rely on dynamic equipment requiring power sources such as pumps were studied worldwide. It was thought that one of the advantages of natural circulation BWRs is that they have a large amount of water in the reactor pressure vessel. GE has compiled the design of a large natural circulation BWR called ESBWR and obtained design certification from the U.S. Nuclear Regulatory Commission in 2011. Based on this design, GE/Hitachi (GEH) Nuclear Energy is proceeding with the approval of a small modular reactor BWRX-300, which arranges small output reactors.

The Oyster Creek Nuclear Power Plant (completed in 1969, electrical output 650 MW) was announced by GE in 1963 as a low-cost boiling water nuclear power plant capable of competing with thermal power generation. Since then, many boiling water nuclear power plants, such as Nine Mile Point and Dresden Units 2 and 3, have been built in the United States and elsewhere. WH also offered pressurized water light water reactors, and orders for light water reactors followed not only in the United States but also worldwide. Oyster Creek has three times the electrical output of Dresden Unit 1. It is worth noting that this contributes to lowering the cost of power generation per kWh. The Tsuruga Unit 1 of Japan Atomic Power Company (1970, electrical output 357 MW) was built with Oyster Creek as a reference plant. Dresden 2 is the first plant of BWR-3, reducing the number of recirculation pipes and installing jet pumps inside the nuclear reactor pressure vessel to reduce construction costs. The Fukushima Daiichi Nuclear Power Station Unit 1 of Tokyo Electric Power Company was built with Dresden Unit 2 as a reference plant. Oyster Creek has been in operation until 2018.

The success of the commercialization of light water reactors in the United States in the 1960s spread worldwide, and Westinghouse and General Electric each built light water reactors by concluding license agreements with power equipment manufacturers in their respective countries. In this way, American light water reactors were exported, and pressurized water light water reactors started operation in Belgium in 1962, Italy and France in 1965, Germany, Spain, and Switzerland in 1969, Japan in 1970, Sweden in 1975, South Korea in 1978, Slovenia in 1983, Taiwan in 1984, Brazil in 1985, the UK in 1995, and China in 2018. Boiling water light water reactors started operation in Italy in 1964, the Netherlands and India in 1969, Japan in 1970, Spain in 1971, Switzerland in 1972, Taiwan in 1978, and Mexico in 1990. In the former Soviet Union, a test reactor of a boiling water light water reactor called VK-50 was built, but it was not commercialized.

The licensing agreement, through its conclusion, enables the use and utilization of intellectual property, making the manufacture and sale of the product possible, and is widely practiced not only in the field of nuclear power generation. If manufactured based on the design of nuclear power plants operating in the United States, it has been confirmed to work well, reducing development risk and construction investment risk. The first unit was built with a U.S. company as the main contractor, and from the second unit onwards, the proportion of manufacturing by the country's manufacturing companies was increased. In this way, U.S. light water reactors have spread worldwide. Japan also promoted domestic production, but when it developed the Japanese-style light water reactor in the 1980s, it sought participation from GE in the development of ABWR. It is said that this was to incorporate good technology from around the world and that Japanese power companies wanted it, but this is unique in the world. For example, West Germany started operating a light water reactor made with a U.S. license in 1969, but terminated the licensing agreement with Westinghouse in a short period of time and started operating a pressurized water reactor of its own design exported to the Netherlands in 1973.

Pressurized water reactors and boiling water reactors each have their own characteristics. The pressure vessel of a pressurized water reactor is smaller than that of a boiling water reactor, but because a steam generator is required, the total weight of these devices is larger than that of a boiling water reactor. However, the secondary system (turbine generator system) of the pressurized water reactor is separated from the reactor cooling water system by the steam generator, so it does not require consideration for radiation protection, and equipment of the same quality as thermal power generation can be used. Boiling water reactors generate electricity by sending the steam produced in the reactor to the turbine generator building. Oxygen, a component of reactor cooling water, absorbs neutrons in the reactor core and becomes radioactive, producing radioactive nitrogen-16. Although the half-life of nitrogen-16 is short, about 7 s, it emits strong gamma rays, so the steam sent to the steam turbine is radioactive. Therefore, radiation shielding is required for the steam turbine and main steam piping of the boiling water reactor, and the quality control requirements for the piping system are stricter than those for the pressurized water reactor.

Regarding the corrosion of equipment by nuclear reactor cooling water, it is difficult to rank the pressurized water reactor and the boiling water reactor. In the pressurized water reactor, corrosion occurs in the thin tubes of the steam generator (heat exchanger), while in the boiling water reactor, it occurs in the reactor core shroud and other internal structures of the reactor pressure vessel. Both types of reactors circulate the reactor cooling water at high temperatures and pressures, making it difficult to completely prevent corrosion through water quality management. If corrosion progresses, these devices and structures are replaced with new ones. Thermal power generation, on the other hand, uses a through-flow cooling system due to supercritical pressure, where all the cooling water is sent to the turbine and fully purified after becoming low-temperature, low-pressure return water. Both types of light water reactors circulate the reactor cooling water at high temperatures and cannot fully purify all of it, making it difficult to deal with stress corrosion cracking of steam generator tubes and internal structures.

In terms of safety, both the pressurized water reactor and the boiling water reactor require power for the emergency cooling pump when the normal power supply is lost. The reactor pressure vessel of the boiling water reactor has the advantage of being larger and holding more water than the pressurized water reactor. On the other hand, the pressurized water reactor can remove decay heat after shutdown using the steam generator. According to the results of the probabilistic risk assessment (NUREG1150) of five types of light water reactors in the United States, the probability of core meltdown in pressurized water reactors is slightly higher than in boiling water reactors. This is thought to be because the cooling system of the pressurized water reactor is an indirect cycle and is more complex than that of the boiling water reactor, but the difference is not significant, and the probability of core meltdown does not necessarily represent the environmental impact in the event of an accident, so the safety is almost equivalent. In conclusion, both the pressurized water reactor and the boiling water reactor have their own characteristics, and it is difficult to rank them.

References

Allen W (1977) Nuclear reactors for generating electricity: U.S. Development from 1946 to 1963, R-2116-NSF, June 1977, The Rand Corporation. https://www.rand.org/content/dam/rand/pubs/reports/2007/R2116.pdf. Accessed August 12, 2024

ANL (2023) Reactors designed by argonne national laboratory, Argonne National Laboratory. https://www.ne.anl.gov/About/ANL-Reactors.shtml. Accessed August 12, 2024

DOE (2015) Atomic power in space II, Department of Energy, INL/EXT-15–34409, September 2015. https://inldigitallibrary.inl.gov/sites/sti/sti/Sort_63995.pdf. Accessed August 12, 2024

EIA (2021) Table 2. Nuclear power plant data as of December 31, 2017, U.S. Energy Information Administration, release date March 20, 2021. https://www.eia.gov/nuclear/spent_fuel/ussnftab2.php. Accessed August 12, 2024

Gavrilas M, Hejzlar P et al (1995) 1995. CRC Press, Safety features of operating light water reactors of western design

Hoshide A, Maekawa O (2008) The evolution of light water reactor plants over half a century, 6th improvement of light water reactors, from Dresden to Tsuruga Unit 1. J Atlc Energy Soc Jpn 50(3):35–39 (in Japanese). https://www.jstage.jst.go.jp/article/jaesjb/50/3/50_169/_pdf/-char/ja. Accessed August 12, 2024

Rahn FJ, Adamantiades AG, et al (1984) A guide to nuclear power technology. John Wiley & Sons, Inc.

Shimada S (2008) The evolution of light water reactor plants over half a century, the 5th improvement of light water reactors, from Shippingport to Mihama Unit 1. J Atlc Energy Soc Jpn 50(2):41–45 (in Japanese). https://www.jstage.jst.go.jp/article/jaesjb/50/2/50_109/_pdf/-char/ja. Accessed August 12, 2024

Simpson JW (1994) Nuclear power from underseas to outer space, American Nuclear Society

Stacy SM (2000) Proving the principle, a history of the Idaho national engineering and environmental laboratory 1949–1999", DOE/ID-10799. https://webharvest.gov/peth04/20041017121453/www.inel.gov/proving-the-principle/. Accessed August 12, 2024

Tong LS (1988) Principles of design improvement for light water reactors, Hemisphere Publishing Corporation

Touran N (2020) History of nuclear reactor development, January 12, 2020. https://whatisnuclear.com/reactor_history.html. Accessed August 12, 2024

Chapter 2
Power Uprating of Light Water Reactors

The power output of many light water reactors around the world has increased over what they were built at. Here explains how.

Nuclear power plants operate with a license obtained after passing a safety review, against 100% thermal output (rated thermal output). In the design submitted for the safety review, the maximum thermal output is determined with design and measurement margins in mind. Once operation begins and actual measurements are obtained, these margins become quantitatively clear. Based on these margins, licenses to operate at higher outputs have been obtained in light water reactors in the United States and Europe. This is called power uprating. Furthermore, by replacing equipment that is limiting power generation, such as power turbines, with larger capacity ones, it becomes possible to operate at higher outputs. This is also called power uprating. In the United States, despite no new light water reactors being built since the 1990s due to the effects of voluntary safety improvements, power generation has increased by about 50%. This increase has been achieved through a reduction in the frequency of power plant shutdowns and an improvement in operation rate (capacity factor) due to the effects of voluntary safety improvements, as well as through power uprating.

Power uprating in the United States has been carried out since the 1970s, but it was systematically carried out from the 1990s to the 2010s. There are three types of power uprating (NRC 2020; Okamoto 2008).

(1) Those that utilize measurement errors: For example, by measuring the cooling water flow more accurately, the accuracy of the reactor output calculated from it is improved, the error is reduced, and the output is increased (called MU type power uprating). Power uprating is within 2%.
(2) Those that increase the output within the range of the initial design (called stretch type power uprating): Those that increase the output within the range of the margin of the initial design without changing the plant equipment. Power uprating is within 7%.

Y. Oka, *Nuclear Power Reactor Development*,
https://doi.org/10.1007/978-981-96-2731-8_2

(3) Those that replace the initial equipment, such as high-pressure turbines, condensate pumps and motors, generators, and transformers, to increase the output (called extended type power uprating). Power uprating is within 20%.

The U.S. NRC has created and published standards for reviewing power uprating applications. Records of power uprating applications are also published. As of 2021, 171 applications have been approved. Many nuclear power plants in the United States have undergone power uprating. From the 1970s to the 1990s, there were many stretch type power uprating, and, since the 2000s, there have been many MU type and extended type power uprating that utilize measurement errors. It is thought that the reason why there are many extended type power uprating since the 2000s is that many power plants perform power uprating when they replace steam turbines and generators of nuclear power plants, which are replaced after long-term operation.

References

NRC (2020). Types of power uprates, U.S.NRC, March 2020. https://www.nrc.gov/reactors/operating/licensing/power-uprates/type-power.html. Accessed August 12, 2024

Okamoto K (2008) Technical evaluation on reactor power uprating, January 15, 2008, 2nd Atomic Energy Commission Document No. 1 (in Japanese). https://www.aec.go.jp/jicst/NC/iinkai/teirei/siryo2008/siryo02/siryo1.pdf. Accessed August 12, 2024

Chapter 3
Advancement of Light Water Reactor Technology

Nuclear power plants are some of the largest structures created by mankind. There are numerous technologies used in light water reactors, which have been improved and advanced over time. Among these, this chapter describes the improvement of light water reactor fuel, research on high conversion light water reactors, the advancement of instrumentation and control technology, and the numerical analysis methods used in nuclear reactor analysis.

3.1 Light Water Reactor Fuel

The fuel of a nuclear reactor depletes due to nuclear fission of uranium, so it is necessary to replace the old fuel with new fuel every few years. Power companies need to regularly purchase new fuel, and the design and manufacture of light water reactor fuel has become a business in itself. Many nuclear fuel manufacturing companies are affiliated with light water reactor design and manufacturing companies. Figure 3.1 shows the fuel assemblies of boiling water light water reactors and pressurized water light water reactors. Fuel pellets are made by sintering uranium dioxide, which are then filled into a metal tube called a fuel cladding tube, and the top and bottom are sealed to make a fuel rod. These are arranged in a square lattice to form a fuel assembly. The fuel pellets are about 1 cm in diameter and length, but one pellet generates energy equivalent to several months of household electricity consumption.

The number of fuel rods in a fuel assembly has increased from those used in the initial light water reactors, without changing the external dimensions. In pressurized water light water reactors, the number of fuel assemblies has increased from 14 × 14, where 14 fuel rods are arranged in a square lattice, to 15 × 15, 16 × 16, and 17 × 17. Boiling water light water reactors also increased from 7 × 7, which was mainstream in the 1960s, to 8 × 8, 9 × 9, and 10 × 10. This is in response to the demand for high burnup of nuclear fuel, which involves reducing the heat load per unit length

Y. Oka, *Nuclear Power Reactor Development*,
https://doi.org/10.1007/978-981-96-2731-8_3

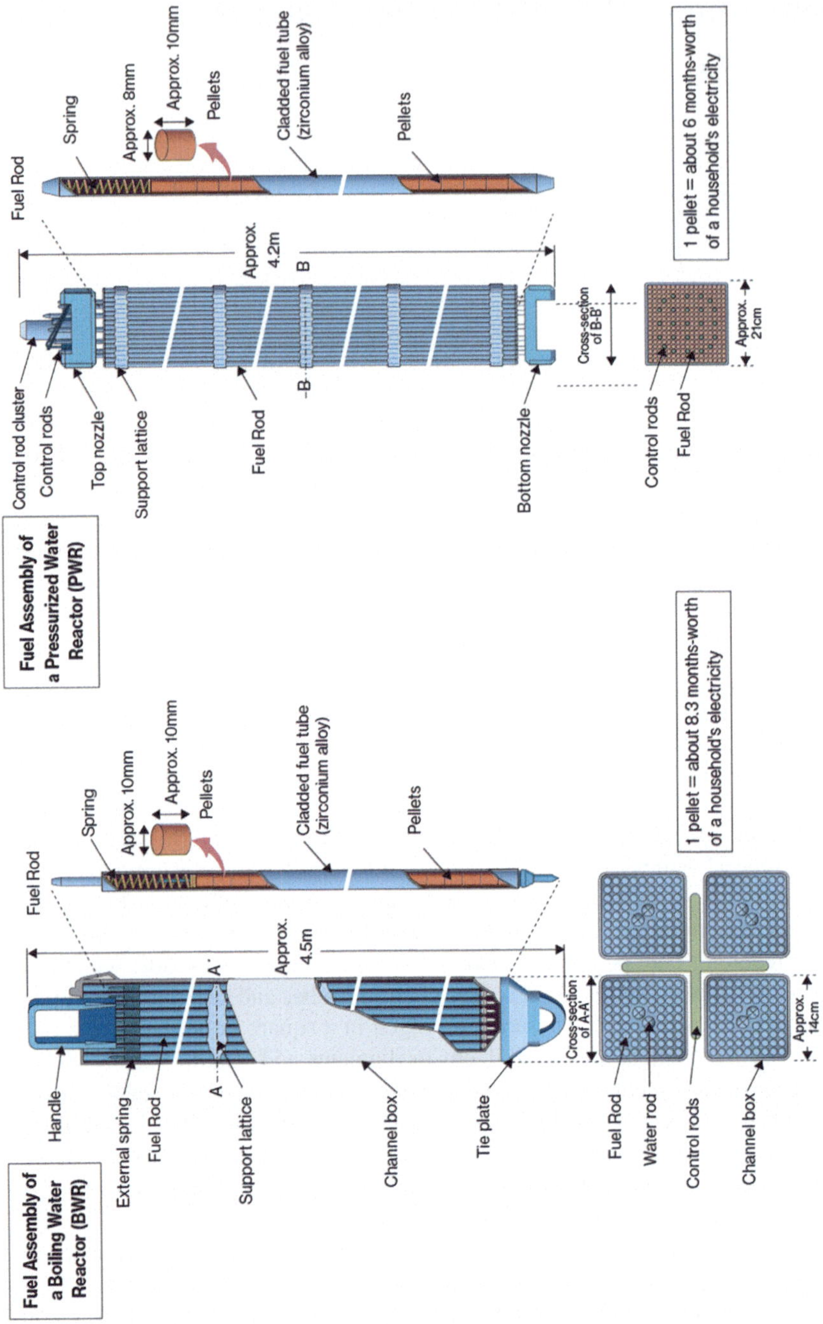

Fig. 3.1 Light water reactor fuel assemblies. *Source* JAERO

of the fuel rod by making the diameter of the fuel rod thinner and increasing the number of rods, in order to use the fuel for as long as possible. The high burnup of the loaded fuel in the reactor, which is used for as long as possible, is a central issue in the improvement of light water reactor fuel, and has contributed to the reduction of fuel costs.

Fuel cladding tubes were made of stainless steel in the early pressurized water reactors, and later of zirconium alloy, while zirconium alloy has been used from the beginning in boiling water reactors. Zirconium is a special metal with low neutron absorption, making it suitable for the cladding tubes of nuclear fuel rods. The drawback is that the chemical reaction of zirconium oxidation by steam at loss of coolant accidents is an exothermic reaction, and the heat from this reaction adds to the decay heat, accelerating the chemical reaction. This chemical reaction reduces water, generating hydrogen. After the Three Mile Island nuclear accident in the United States in 1979, reports reevaluating stainless steel cladding tubes were released from the Electric Power Research Institute in the United States, but zirconium alloy cladding tubes have continued to be used. In recent years, improved fuel cladding tubes have been researched and developed under the name of accident-tolerant fuel.

In the fuel assemblies of boiling water reactors, a cover made of zirconium alloy, called a channel box, is attached and used as shown in Fig. 3.1. A cross-shaped control rod is inserted in the center of the four fuel assemblies. Boiling water reactors create steam in the core through saturated boiling. The lower part of the fuel assembly is water, and the upper part is a boiling two-phase flow. The fuel assemblies of boiling water reactors are removed after approximately four operating cycles (each operating period divided by shutdowns for fuel replacement). Since one of the four fuel assemblies is replaced in each operating cycle, the burnup is different for each. The channel box prevents cross-flow (cooling water entering and exiting each other) between fuel assemblies and keeps the cooling flow rate of each fuel assembly constant. In contrast, the fuel assemblies of pressurized water reactors do not have a channel box, and the cooling water of the nuclear fuel assembly can flow into adjacent fuel assemblies (open lattice structure). The control rod elements are rod-shaped, sturdy, and bundled into control rod clusters, inserted into the control element insertion tubes in the fuel assembly, and inserted and removed to control startup, shutdown, and core power distribution.

3.2 Research on High Conversion Light Water Reactors

In the 1950s, when the development of commercial nuclear power began in the United States, it was thought that uranium resources would be scarce in the future. Therefore, research was conducted on breeder reactors and high conversion reactors. In addition to fast breeder reactors using liquid sodium, research was conducted on cores that produce Uranium-233 from thorium in light water reactors, and cores that use uranium fuel and have a high conversion rate to plutonium (high conversion light water reactors) (Rahn 1984, pp. 474–490).

Thorium is not fissile, but when it absorbs neutrons, it produces fissile Uranium-233. By adding thorium to the fuel, irradiating it, reprocessing the spent fuel to extract Uranium-233, and processing it into fuel, it is possible to breed fuel even in a light water reactor. The reason for this possibility is that Uranium-233 emits a large number of neutrons due to thermal neutron fission. In the Shippingport PWR, tests have been conducted on a thermal neutron breeder reactor core consisting of a Uranium-233 fuel section and a thorium fuel section, achieving a breeding ratio of 1.01. Thorium resources are abundant in places like India, and occasionally, breeder reactors using thorium fuel become a topic of discussion. Thermal neutron breeder reactors have the advantage of being able to use existing light water reactors as they are, except for the fuel, but Uranium-233 does not exist naturally, so initially, enriched uranium or plutonium is needed as fuel. There are issues with the cost of developing and installing reprocessing technology and facilities to extract Uranium-233 from spent fuel. Furthermore, Uranium-233 has a shorter half-life and higher radiation levels than Uranium-235 or Uranium-238, so there are issues such as the need to consider radiation protection when fabricating nuclear fuel. In contrast, uranium fuel consisting of U235 and U238 is easy to handle.

By decreasing the gap between the fuel rods of the fuel assembly of a light water reactor, reducing the moderation of neutrons (Decelerating the speed of neutrons is called "moderation" in nuclear engineering.), and increasing the conversion rate from Uranium-238 to plutonium. Research on the core of a high conversion light water reactor was initially conducted in the United States, and later in Europe and Japan. In the 1980s, research in the United States and Europe declined due to falling uranium prices and concerns about nuclear proliferation. In Japan, research on boiling water type high conversion light water reactors continued for a long time (Uchikawa 2005, 2007). In the case of light water cooling, even if the fuel rod gap is reduced from the usual 3 mm to 1 mm in a boiling water type, the breeding ratio (plutonium surviving ratio) is just over 1.0. In the case of a pressurized water reactor, it is difficult to exceed a breeding ratio of 1 because the water density is higher than in a boiling water type.

3.3 Instrumentation Control of Reactors

In nuclear power plants, not only neutron flux but also temperature, pressure, and coolant flow conditions are measured by sensors, and signals are collected and displayed in the central control room to control the reactor and power generation system. The instrumentation control technology for this has been improved and evolved. The instrumentation control technology of reactors has progressed in step with the transition from analog to digital with the development of electronic computers. Until the 1970s, it was the analog era using relays, switches, and meters, displaying the physical quantity obtained from one detector with one instrument, and monitoring the state of the plant. Since there are many instruments in a nuclear power plant, the central control panel that displays and controls them was also huge. Figure 3.2 shows the evolution of the central control panel in nuclear power plants

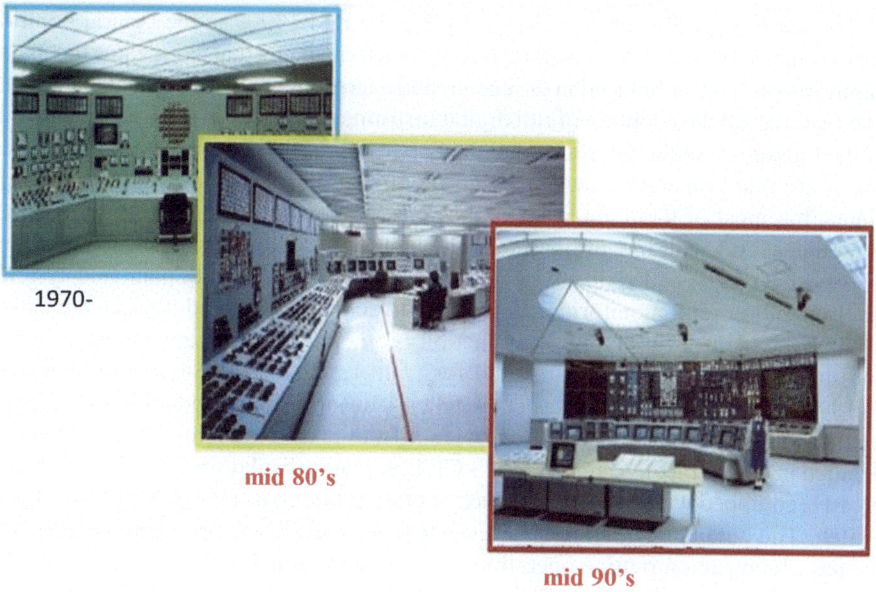

Fig. 3.2 Evolution of central control panels in Japanese nuclear power plants. *Source* TEPCO

in Japan. The control rooms of nuclear power plants in the 1970s were based on the design of control rooms of thermal power plants in the 1960s. It is a control panel from the analog era, with switches, meters, indicator lamps, and other instruments arranged on a bench-type control panel.

In the Three Mile Island nuclear accident that occurred in the United States in 1979, numerous warning displays flashed simultaneously, confusing the operators. The cause of the accident was a combination of factors, including operator error. This incident highlighted the importance of the man–machine interface, and improvements were made to the control panel to display important information to the operators. In the control panels around the mid-1980s, the main and auxiliary panels were separated, color Cathode Ray Tubes (CRTs) were adopted, and partial automatic control was introduced. For important systems such as the emergency core cooling system, they were grouped by multiple divisions, and the positional relationship of pumps and valves was made understandable by mimicking (displaying the functional relationship of equipment according to the flow of the process in a system diagram), making it easier for operators to identify and recognize. Color displays were used to clearly distinguish systems by color. The classification of the importance of monitoring equipment was carried out, and the instrumentation control panel was divided into main and auxiliary panels. The central control panel is designed so that the alarm display device can be confirmed from the rear, the indicator recorder and its display device are located in the middle position so that they can be monitored from the front when standing in front of the control panel, and the operators are provided in the lower section (desk part).

The control panel shown in Fig. 3.2 from the mid-1990s as the central control panel is the ABWR control panel. It consists of a main panel that aggregates the main monitoring operations placed in the center, and a large display board placed behind it. The features are the adoption of full digital instrumentation control, the advancement of the human–machine interface, and the expansion of the automatic driving range. There are many operation switches and monitoring information in nuclear power plants, but most of them are operation switches and monitoring information at the equipment level with low usage frequency. For these, a touch-operated CRT or flat display that can be called up as needed is used. On the other hand, plant and system level operations and monitoring are important, so dedicated fixed devices are used. That is, the plant-level monitoring function is a large display board mimic display, and dedicated hard switches are used for plant-level safety system start switches and system-level automation master switches. The large display board places a large mimic that can grasp the overall overview of the plant in the center, and displays detailed information of the main board's CRT on a large display on the right. The main board is equipped with a CRT with touch operation function, a flat display, and a hard switch. The automatic operation of nuclear power plants has been automated from the past, focusing on routine operations such as start-stop, but the automation range has also been expanded. In ABWR, automation of routine operations immediately after reactor scram (emergency stop) and control rod search is aimed at reducing the burden on operators.

The digitization of the instrumentation control panel was introduced to the peripheral equipment of the nuclear power plant, accumulated experience, and then introduced to the main control panel. In the mid-1980s, first, the control panel of the radioactive waste treatment system of the nuclear power plant was digitized. Since the radioactive waste treatment system is a separate system from the instrumentation control system of the nuclear power plant, even if trouble occurs, it does not affect power generation, and experience in digitization could be accumulated. After that, it was introduced to the non-safety system part of the instrumentation control system of the nuclear power plant, and then introduced to the safety system based on that experience. In ABWR, full digitization has been achieved. Digitization uses a large number of Central Processing Units (CPUs) and software. In order to guarantee the quality of software and digital products for safety systems, Verification and Validation (V&V) methods were determined and used as guidelines of Institute of Electrical and Electronics Engineers (IEEE) and Technical Guidelines of the Japan Electric Association (JEAG).

In thermal power plants, small, distributed control devices that use commercially available computers and software on an open platform (specifications and technologies are disclosed and can be provided by other companies) have been introduced into the instrumentation control system, and control panels operated by a mouse are used. However, in the case of nuclear power plants, it was difficult to use such general-purpose, commercially available products in their instrumentation control systems due to safety requirements. Safety requirements include system separation, redundancy, inspectability, reliability, quality control and quality assurance, compliance with guidelines and standards, and software V&V. Furthermore, in the instrumentation control system of nuclear power generation, it is required that the parts used can be supplied and maintained for a long period of time, that the cause can be identified when an abnormal situation occurs, and that horizontal deployment can be made to related products and parts for problem solving. Therefore, special designs and products are needed for the instrumentation control system of nuclear power plants. However, digitization has been promoted because of its great benefits.

The progress of the central processing unit (CPU) of computers is extremely fast, and what is installed quickly becomes outdated, and the manpower required for the V&V of its software is enormous, so commercially available computers and their software cannot be used in the measurement control system of nuclear power plants for a long period of time. Therefore, a special computational device called Field Programmable Gate Array (FPGA) is used in the instrumentation control of nuclear power plants.

FPGA is a device (integrated circuit) that integrates gates (logic circuits) that can be configured by the purchaser or designer in the field (on-site) after manufacturing. FPGA is a type of Programmable Logic Device (PLD), but while PLD has hundreds of gates, FPGA has thousands of gates, and is used not only for prototyping circuit designs, but also for commercial integrated circuits, mobile phones, automobiles, game devices, cameras, medical electronic devices, and many other fields. When FPGA is used for the digitization of the instrumentation control system of a nuclear power plant, the life of the product does not need to respond to changes in the CPU market. Moreover, by using FPGA, various electronic circuit functions can be constructed with a small variety of FPGAs, and the reliability of design and manufacturing as an integrated circuit can be ensured. This ensures the long-term maintenance of the instrumentation control system of a nuclear power plant, the supply of parts, and the sustainability of design. In terms of V&V, FPGA does not have a CPU or Operating System (OS, basic software of a computer), so it can be simplified. The design of the logic circuit using FPGA is done using a hardware description language called VHDL. VHDL is the same standard in IEEE and International Electrotechnical Commission (IEC). Circuit design, circuit implementation, and hardware manufacturing that were done in analog instrumentation control systems correspond to design by VHDL, its implementation, and hardware manufacturing, respectively. This has realized a digital instrumentation control system that does not require a CPU or OS (Maekawa 2011). It should be noted that digital technology is advancing rapidly, and if responses to standards, regulations, and requirements for instrumentation and control systems in the construction of new nuclear power plants are delayed, construction may be delayed. This has been a factor in construction delays and cost overruns in Europe, so it is necessary to be mindful of this in the future.

The method of controlling nuclear power plants is PID control, which has been widely used in the control of engineering equipment since ancient times. PID control is a type of feedback control, and it controls the input value by the deviation of the output value and the target value proportional (P, proportional), integral of the deviation (I, integral), and differential (D, differential). It is a method that has been systematized within the framework of classical control theory. It is said to have developed with the development of automatic steering of ships in the early twentieth century. Because of its good compatibility with intuition based on past performance and the experience of engineers, it is still used as a main control method in the industrial world, not limited to nuclear power generation.

3.4 Nuclear Reactor Analysis Technology

In the field of nuclear power, numerical analysis using electronic computers has been conducted since the early stages of its development. The national research institutes that conducted the early nuclear power development in the United States used newly developed electronic computers for nuclear weapon development. Nuclear power development was a major user that promoted the development of electronic computers, and it is no exaggeration to say that the early electronic computers developed along with nuclear power development. The national research institutes under the U.S. Department of Energy, which conducted nuclear power development, still install and use the most advanced computers. In Japan, in the 1960s, the most advanced large-scale electronic computers were installed in the computing center of the Japan Atomic Energy Research Institute (now the Japan Atomic Energy Agency) and used for research and development. In the private sector, a business started that installed and used large-scale computers made in the United States, and they were used for calculations for the development of power reactors.

Nuclear power plants are the largest artificial objects created by mankind. The only thing that corresponds to the construction cost of one nuclear power plant in Japan is the Akashi Strait Bridge, which costs about 500 billion yen in total construction cost. The fields involved in the design and construction of nuclear power plants span most of the fields of engineering. Nuclear engineering is a comprehensive engineering, consisting of nuclear reactor physics, thermal fluid engineering, structural mechanics, materials science, applied chemistry, etc. in the field of engineering. There are textbooks for each field, but the seeds of development are often found at the boundaries between each field or outside, so it is good to understand the overview of analysis technology related to the design and safety of nuclear power plants in an overview. Here, we will describe the overview of nuclear reactor core design and the dynamic characteristics and safety analysis of nuclear power plants. For details on individual physical phenomena and analysis methods, please refer to the respective specialized books. **After learning all the analysis methods, reading the following overview should help you get an overview of the whole.** Before discussing the method of nuclear reactor analysis, let's talk about the basic physical phenomena of a nuclear reactor, namely the nuclear fission chain reaction and the moderation (deceleration) of neutrons.

3.4.1 Nuclear Fission Chain Reaction and Neutron Moderation

In 1942, Enrico Fermi was the first in the world to achieve a nuclear fission chain reaction using a pile (nuclear reactor) in which natural uranium fuel rods were inserted into a stack of graphite in a laboratory set up under the bleachers of the University of Chicago's football field. When a neutron collides with a uranium nucleus (Uranium-235) in the fuel rod, a nuclear fission reaction occurs, and two or more neutrons are generated. One of these causes the remaining uranium nuclei to undergo fission, resulting in a continuous series of fissions via neutrons. This is the nuclear fission chain reaction. In a nuclear reactor operating at a constant output, nuclear fission chain reactions occur at a constant rate. This is referred to as the reactor being in a critical state. Both a 1-W reactor and a 1-million-watt reactor are in a critical state because their output is maintained at a constant value. To stop a reactor, control rods containing neutron-absorbing material are inserted into the reactor core to make it subcritical. In a subcritical state, the number of neutrons disappearing is greater than the number being generated, so the fission rate (output) decreases. To increase the output, the control rods are withdrawn to reduce the neutron absorption rate, slightly supercritical until the target output is reached, and then returned to critical. When the number of nuclear fission chain reactions is decreasing or the reactor is stopped, the reactor is subcritical. A state in which the output is increasing is supercritical. There are two types of supercritical states: delayed supercritical, where the output increase rate is small, and prompt supercritical, where the output increases rapidly. Nuclear power plants generate electricity by keeping the reactor in a critical state. While thermal power plants control output by varying the amount of fuel supplied, nuclear power plants vary output by controlling the nuclear fission chain reaction. Reactors need to be designed to have a negative power reactivity coefficient. That is, when a positive reactivity is added and the output increases, the reactor is designed to have an inherent characteristic that suppresses the increase in output.

The energy generated by the movement of atomic nuclei in a nuclear fission reaction, or the energy generated when an unstable atomic nucleus decays to a stable one, is converted into heat in the reactor fuel. This heat is transferred to the reactor coolant flowing outside the fuel rods, which generates steam, and the steam rotates the steam turbine generator to generate electricity. Power reactors utilize the energy produced by nuclear fission chain reactions.

When the energy (flight speed) of neutrons decreases, they are more likely to react with uranium nuclei. For this reason, Fermi used graphite. Neutrons produced in nuclear fission repeatedly collided with graphite (carbon atom), reducing their energy and causing nuclear fission reactions more frequently. In addition to graphite, heavy water and light water can also be used to reduce the energy of neutrons. Because the energy (flight speed) of neutrons decreases, this is called moderation (deceleration). Of graphite, heavy water, and light water, light water is the most effective at moderating (decelerating) neutrons. The reason for this is that the mass of the hydrogen nucleus that makes up light water is the same as that of the neutron.

The mass numbers of the atomic nuclei of graphite, deuterium, and hydrogen are 12, 2, and 1, respectively. The mass number of a neutron is 1. When a neutron collides with a hydrogen nucleus, both masses are the same, so if it is a head-on collision, the energy (flight speed) of the neutron becomes zero in one collision, like when a billiard ball collides head-on with a stationary ball. However, when a neutron collides with the atomic nucleus of graphite or deuterium, the opponent's atomic nucleus is heavier, so it bounces back, and the energy lost in one collision is small, and the energy of the neutron does not decrease unless it collides many times. Neutrons whose energy has sufficiently decreased and no longer decreases are called "thermal neutrons". Since atomic nuclei are in thermal motion, when the energy of neutrons is low, they may gain energy when they collide with atomic nuclei. The energy of thermal neutrons is not zero, but corresponds to the thermal equilibrium state related to the temperature of the moderator. The energy of neutrons produced in nuclear fission is about eight orders of magnitude larger than the thermal neutrons.

Thermal neutrons have a high probability of causing a nuclear fission reaction when they collide with uranium, and low-enriched uranium or natural uranium can be used as nuclear fuel. Light water absorbs a larger proportion of neutrons when they collide compared to heavy water or graphite, so natural uranium fuel cannot maintain a nuclear fission chain reaction, and the nuclear fuel for light water reactors is low-enriched uranium. Fermi was able to realize a nuclear fission chain reaction using natural uranium as fuel because he used graphite. Graphite, heavy water, and light water are called moderators. A reactor that uses a moderator to cause a nuclear fission chain reaction is called a thermal neutron reactor. A reactor that causes a nuclear fission chain reaction with fast neutrons produced in nuclear fission without using a moderator is called a fast reactor. Fast reactors need to use plutonium or highly enriched uranium as fuel. Light water reactors, graphite-moderated gas-cooled reactors, and heavy water-moderated light water-cooled reactors are thermal neutron reactors.

3.4.2 Nuclear Data and Neutron Cross-Section

Nuclear power generation utilizes the reaction between neutrons and atomic nuclei. The foundation of its analysis is nuclear data. Nuclear data is a general term for data not only on the interaction between neutrons and matter, but also on the interaction between protons, alpha particles, etc. and atomic nuclei, the excited and stable states of atomic nuclei, the probability of their transitions, and the radiation they emit or absorb. Nuclear data is primarily determined based on experimental values, but measurement data contains various errors. Moreover, it is not possible to measure data for all energy points, reactions, and nuclear species. On the other hand, the nuclear data required for reactor analysis must be given as a set of continuous values within the energy range being handled. Therefore, work is being done to find the nuclear data that is thought to be closest to the true value by adding calculated values based on nuclear models created based on the theory of nuclear physics and estimated values using statistics. This is called nuclear data evaluation. The evaluated nuclear data is compiled into a data file according to a predetermined format, which is called an evaluated nuclear data library. The evaluated data is processed in various ways and used according to the purpose. In the design and analysis of nuclear reactors, the energy range of the neutrons being handled spans 10 digits, so it is common to use the neutron cross-section (the likelihood of a reaction between a neutron and an atomic nucleus) as a group constant averaged for each type of reaction in each energy interval. The evaluated nuclear data library includes the Evaluated Nuclear Data File (ENDF) developed mainly in the United States, Japan's Japanese Evaluated Nuclear Data Library (JENDL), and Europe's Joint Evaluated European Fission and Fusion File (JEFF; predecessor is JEF, Joint European File). An example of the evaluation of the neutron cross-section of Scandium (Sc-45) is shown for reference (Oka 1982). The nuclear data evaluation results are compiled by comparing calculations using the nuclear model and experimental values.

3.4.3 Core Nuclear Calculation Method

In physics, the laws of nature are formulated and used as equations. The equation that describes the behavior of neutrons in matter such as a nuclear reactor is Boltzmann's transport equation. In physics, Boltzmann's transport equation is known as an equation that describes the motion of gas molecules. In the case of the kinetic theory of gases, it is necessary to consider the collision of gas molecules with each other, so it is a nonlinear equation, but in the case of reactor analysis, the collision between neutrons can be ignored compared to the collision between neutrons and the atomic nuclei of the reactor, so the neutron transport equation used in reactor analysis is a linear equation. The neutron transport equation has seven independent variables: space (3 dimensions), energy (1 dimension), direction (2 dimensions), and time (1 dimension). When neutrons flow in a certain direction, etc. and their anisotropy is strong, it is necessary to use the neutron transport equation: For example, the analysis of neutron transmission behavior in radiation shields placed around a nuclear reactor, or the analysis of neutron behavior in the blanket section placed around the plasma of a fusion reactor.

The behavior of neutrons in the core of a nuclear reactor, excluding the peripheral parts, does not have strong anisotropy, so it can be approximately represented by a neutron diffusion equation that does not have a dimension representing direction. The diffusion equation is an equation that represents, for example, how the concentration of ink molecules dropped into water dilutes as they repeatedly collide with water molecules. Neutrons also spread in the reactor core by repeatedly colliding with the atomic nuclei that make it up. The neutron diffusion equation is a five-dimensional equation that describes how neutrons spread from areas of high density to lower ones. In a steady state (a state where the reactor is operated at a constant output), the dimension of time disappears, so it becomes a four-dimensional equation with spatial and energy variables.

Solutions to the neutron transport equation and the neutron diffusion equation can be obtained analytically in the case of simple systems and boundary conditions. In reactor analysis, the distribution of neutron number density in the reactor is determined by numerical calculation. (Digital) computers cannot perform analytical calculations such as differentiation and integration, but they can solve large-scale algebraic equations. The neutron number density and related functions represented by the neutron transport equation and the neutron diffusion equation are rewritten into algebraic equations by discretizing them. That is, the function of continuous variables is replaced with a set of values at discrete points. This allows us to determine the distribution of neutron number density in complex systems. This method is widely used in numerical calculations, not limited to neutron transport and diffusion equations. The output distribution of the reactor can be determined from the obtained neutron number density distribution (Duderstadt 1976).

Nuclear fission occurring in a reactor, and the scattering and absorption of neutrons, are probabilistic phenomena governed by occurrence probabilities represented by neutron cross-sections. The Monte Carlo method is a method of obtaining a solution by calculating the behavior of a single neutron according to the probability represented by the neutron cross-section using random numbers, and summarizing the calculation results (trial results) of many neutrons. A large number of trials are necessary to obtain a significant solution. With the improvement of computer capabilities, nuclear calculations by the Monte Carlo method have also been put into practical use. The continuous energy Monte Carlo method, which does not use the neutron cross-section discretized as group constants, was developed at Los Alamos National Laboratory in the United States in the late 1980s and is used for radiation shielding calculations and detailed calculations of reactor static characteristics.

3.4.4 *Reactor Burnup Calculation*

The fissile material (nuclear fuel material that can undergo nuclear fission when struck by a neutron of low energy) such as U-235 in a reactor decreases due to nuclear fission caused by neutrons. The fertile material (a material that, although not fissile itself, can be converted into a fissile material by neutron absorption) such as U-238 contained in the nuclear fuel capture neutrons and become different atomic nuclei. These atomic nuclei undergo transitions that emit beta decay and gamma rays to produce plutonium, an artificial atomic nucleus. The time changes of each of these atomic nuclei are governed by the generation rate due to decay, the generation rate due to neutron capture, the extinction rate due to neutron absorption, and the extinction rate due to decay. Solutions can be obtained by solving the equations representing the time changes of each atomic nucleus. There are several numerical solution methods for these simultaneous equations, such as the Bateman method.

3.4.5 Reactor Kinetics Calculation

The time change characteristics of reactor output are called reactor kinetics. The time change characteristics referred to as reactor kinetics are not about the long-term changes in reactor characteristics due to fuel consumption, but about the output change characteristics in seconds to several minutes during reactor startup, shutdown, output change, accidents, etc. The point reactor kinetics equation, which ignores the spatial and energy dependencies of the neutron number density distribution, is used for the analysis of reactor kinetics. There are very few types of unstable atomic nuclei produced by nuclear fission that emit neutrons and transition to a stable state. While fission neutrons are emitted simultaneously with fission, the neutrons generated by the decay of unstable atomic nuclei are not emitted simultaneously with fission but are delayed, so they are called delayed neutrons. Criticality is a state where neutron generation and extinction are balanced, and the reactor output is constant. If the balance is disturbed, the power output changes. Therefore, it is necessary to consider delayed neutrons in reactor kinetics. The equation for delayed neutrons describes the generation and extinction of atomic nuclei that produce delayed neutrons, divided into six groups according to their half-lives of the unstable atomic nuclei. The point reactor kinetics equation is a system of seven equations consisting of a first-order time-dependent differential equation that shows the balance of generation and extinction of fission neutrons, and six first-order time-dependent differential equations that describe the generation and extinction of delayed neutrons. Solutions to these equations can be obtained by Laplace transformation or Fourier transformation.

In actual power reactors, changes in the temperature of nuclear fuel and the density of cooling water due to changes in reactor output change the neutron absorption rate, etc. causing the fission rate to change and the output to change. This is called output feedback. In this case, it is necessary to consider this effect in the reactor kinetics equation. Furthermore, since the reactor output is also controlled by the control system, plant dynamic characteristic analysis that takes these into account is necessary.

3.4.6 Thermal Hydraulic Calculation of Reactor Core

The reactor core is composed of many fuel rods. In the reactor core, the heat generated by the fuel rods is removed by the coolant flowing around it. The basic model of thermal hydraulic calculation in the reactor core is the single-channel model which is shown in Fig. 3.3. It is a one-dimensional cylindrical shape of a single fuel rod and its equivalent coolant flow path around it. The fuel rod consists of a long hollow tube called a metal cladding tube, in which fuel pellets made by baking nuclear fuel material are packed and the tube is sealed at the top and bottom. The heat generated by the fuel pellets is transmitted to the coolant, resulting in a rise in the temperature of the coolant.

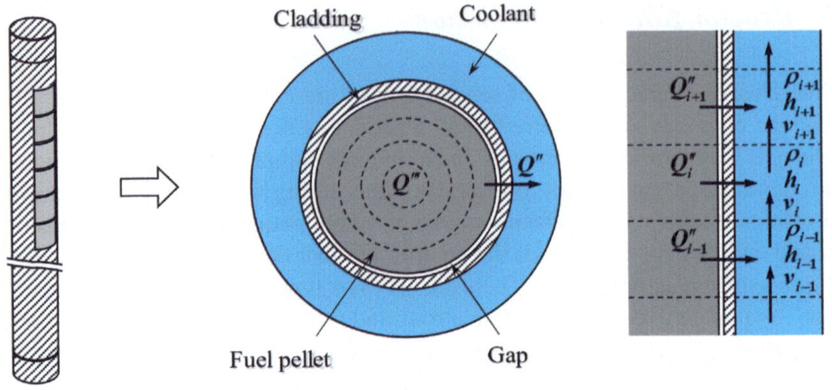

Fig. 3.3 Single channel heat transfer model of a fuel rod

In the single-channel model, the heat flux Q" generated by the fuel rod is transmitted radially to the surrounding coolant. In the axial direction, the coolant flows from bottom to top, and the heat of Q" per unit area is transmitted to the coolant. The state quantities of the coolant (density ρ, enthalpy h, flow rate υ) change due to the transmitted heat. By giving the state quantities such as the inlet temperature (enthalpy) and flow rate of the coolant, and calculating the change in state quantities in accordance with the energy conservation law and the mass conservation law from the bottom node, the outlet temperature of the coolant can be determined. If this is calculated for the average fuel rod in the reactor core, the outlet temperature of the reactor coolant can be determined (Oka 2014). The coolant of the fuel assembly bundled with fuel rods is mixed with the coolant around the adjacent fuel rods. This heat transfer flow analysis is called subchannel analysis and is performed. In this case, pressure loss and momentum conservation law associated with heat transfer flow are also considered in the calculation.

3.4.7 Nuclear-Thermal Coupling Calculation: Calculation of Core Output Distribution

In reactors such as light water reactors where cooling water also serves as a neutron moderator, changes in cooling water temperature or boiling cause changes in the density of the moderator, changes in the degree of moderation, and changes in output. The output distribution also changes in response to changes in moderator density. To determine these core characteristics, the heat generation distribution of the core obtained by nuclear calculation is input, and a large number of single-channel calculations corresponding to each heat generation distribution are performed for the entire core to determine the density distribution of the coolant/moderator. Based on this, the neutron cross-section used in the core nuclear calculation is changed, and the nuclear calculation is performed again. This is repeated until the calculation converges. This is called a nuclear-thermal coupling calculation. Boiling water type light water reactors experience saturated boiling in the core, and the density of the cooling water changes significantly, so nuclear-thermal coupling calculation is essential. Pressurized water reactors experience nuclear boiling at the top of the core. The change in cooling water density is not as large as in boiling water type light water reactors, but nuclear-thermal coupling core calculations are performed to accurately determine the core output distribution.

3.4.8 Thermal Hydraulic Calculation of Nuclear Power Plants: Safety Analysis

The simplest and most basic thermal fluid calculation model used in the safety analysis of nuclear power plants is the node junction model. An example of this is shown in Fig. 3.4. This figure is for the plant dynamic characteristic analysis of a supercritical pressure light water reactor, but for safety analysis, a model that adds a low-pressure injection system and a pressure suppression pool to the node junction model is used. For boiling water reactors and pressurized water reactors, models that add each system's equipment, such as recirculation pumps and their cooling water circulation pipes, steam generators and their heat transfer fluid pipes, and for safety analysis, emergency core cooling system equipment, pipes, and valves are used. The calculation is done in the order of the flow of the cooling water, from the upstream node (section for calculation) according to the mass conservation law and the energy conservation law. The steam table is a numerical table that shows the thermodynamic properties of water. It includes the specific enthalpy and density of water and steam as parameters of pressure and temperature, and are referred to in calculations.

There are various types of accidents and abnormal transients in nuclear power plants. Among the accidents, those like the loss of coolant accident, where the output of the reactor automatically decreases due to the negative feedback characteristics inherent in the reactor, and the control rods are inserted and the reactor stops, do not require calculation of the reactor dynamics as the time change of the output decrease can be given as input in the calculation. In this case, the accident analysis becomes a thermal hydraulic calculation. Even in cases like a reactor control rod pull-out accident, the reactor automatically stops (scrams), so the time change of the reactor output decrease after that can be given as input, so the accident analysis becomes a thermal hydraulic calculation. The calculation program (calculation code) to calculate the accidents and abnormal transients of a nuclear power plant started from a simple node junction model, and various improvements such as consideration of pressure loss and momentum conservation law have been made. In the United States, there are safety analysis codes and reactor calculation codes developed by

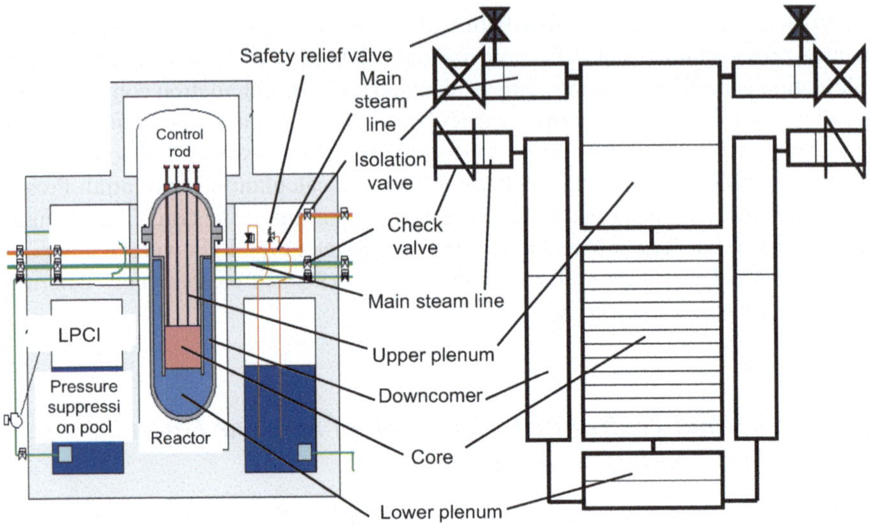

Fig. 3.4 Node junction model

the Nuclear Regulatory Commission, the nuclear industry, and national research institutes. The U.S. Regulatory Commission introduces the analysis codes on its website as thermal hydraulic calculation codes (NRC 2023).

3.4.9 Plant Dynamic Calculation and Plant Control

If the reactor does not stop and the output change continues, plant dynamic characteristic analysis is necessary. It is necessary to consider the effect of control by the nuclear control system. An example of the plant control system is shown in Fig. 3.5.

This is an example of a supercritical pressure light water reactor. In this example, the output of the reactor is controlled by control rods, the pressure is controlled by the main steam regulating valve of the steam turbine, and the control of the reactor feedwater temperature is performed by the main feedwater pump. What to control with is determined by examining the sensitivity to changes in the control amount. Since it is better to control with something that has a high sensitivity, the control method is not unique. In nuclear power plants, empirically determined control methods are used for each. For example, in boiling water light water reactors, a recirculation flow control method that changes the output by controlling the recirculation flow and changing the amount of bubbles (void fraction) caused by boiling in the core is used together with output control by control rods. The principle of control is Proportional, Integral, Differential (PID) control, which is described in the section on instrumentation control.

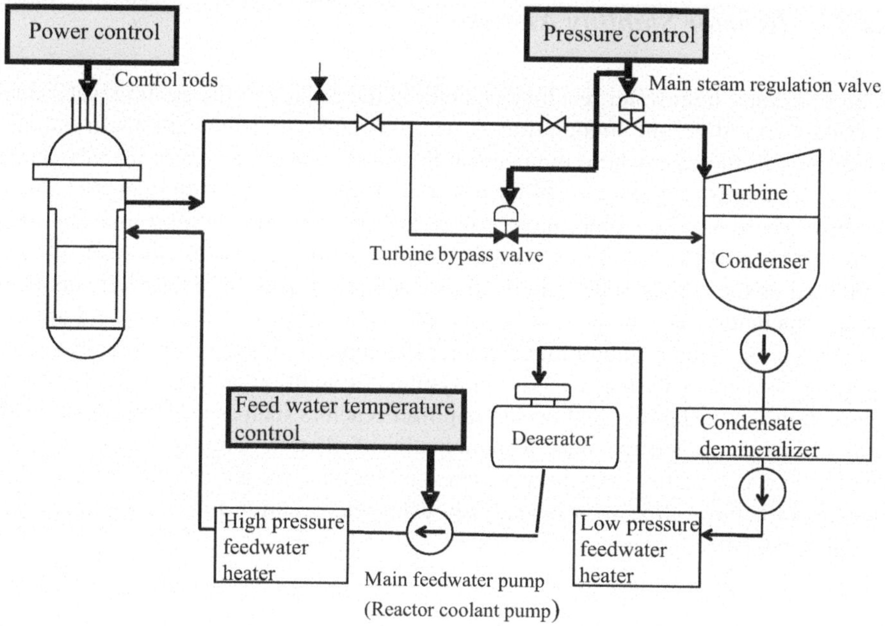

Fig. 3.5 Plant control system

3.4.10 Plant Startup

In nuclear power plants, it takes several tens of seconds for the reactor cooling water to come out of the reactor, go through the steam turbine, condenser, deaerator, and feedwater heater, and return to the reactor. When analyzing time changes that take longer than this, for example, control including startup and shutdown, and plant stability analysis of boiling water light water reactors, it is necessary to include systems called balance of plant such as steam turbine systems, condensers, feedwater heaters, and pumps in the calculation model. The balance of plant does not need to be modeled in detail; it is sufficient if the conservation of mass and energy and the time behavior of the plant are simply expressed. The plant startup procedure is determined to meet the constraints at startup. In light water reactors, in addition to the nuclear startup conditions to achieve criticality, there are restrictions on heat flux, the temperature rise rate of the reactor pressure vessel due to the thermal stress, and various stabilities and restrictions on the moisture content of the steam driving the turbine.

3.4.11 Reactor Stability Analysis

In boiling water light water reactors, changes in the density of the cooling water due to boiling can affect the moderation of neutrons and cause fluctuations in output. In addition to the case where the neutron flux in the core changes uniformly (core stability), there is regional instability where it changes non-uniformly, and channel stability where a single fuel channel flow vibrates. In this stability analysis, the state variables of the equations used for plant dynamic characteristic analysis are expressed as the steady-state value and its fluctuation, and the linearized equation for the fluctuation is obtained and analyzed in the frequency domain by Laplace transformation. This is called linear stability analysis. Plant stability is analyzed in the time domain including the control system. In addition to this, there is xenon spatial oscillation stability that occurs in power reactors such as graphite-moderated carbon dioxide-cooled reactors. Xenon spatial oscillation stability occurs in relation to the accumulation and disappearance of xenon, a fission product, and spatial changes in neutron flux, so it is analyzed using the equation for xenon generation and disappearance.

References

Duderstadt JJ, Hamilton LJ (1976) Nuclear reactor analysis. John Wiley and Sons Inc. New York, 1976. https://deepblue.lib.umich.edu/bitstream/handle/2027.42/89079/1976_Nuclear_Reactor_Analysis.pdf. Accessed August 12, 2024

Maekawa M, Hayashi T (2011) Chapter 8, next generation technologies in the digital I&C systems for nuclear power plants. In: Y Oka, et al (ed) Advances in light water reactor technologies, Springer. https://link.springer.com/chapter/10.1007/978-1-4419-7101-2_8. Accessed August 12, 2024

NRC (2023) Computer codes, US NRC, last updated March 10, 2023. https://www.nrc.gov/about-nrc/regulatory/research/safetycodes.html. Accessed August 12, 2024

Oka Y, Nakagawa T, Kikuchi Y (1982) Evaluation of neutron nuclear data for scandium-45, JAERI-M 9981, February 1982 (in Japanese). https://jopss.jaea.go.jp/pdfdata/JAERI-M-9981.pdf. Accessed August 12, 2024

Oka Y (ed) (2014a) Nuclear reactor design, Section 2.2, Springer, 2014. https://link.springer.com/book/10.1007/978-4-431-54898-0. Accessed August 12, 2024

Rahn FJ, Adamantiades AG, et al (1984) A guide to nuclear power technology. John Wiley & Sons, Inc.

Uchikawa S (2005) Innovative water-cooled reactor: current status and prospects of reduced-moderation water reactor research and development, JAERI-Conf 2005–009, August 2005, pp 59–84 (in Japanese). https://jopss.jaea.go.jp/pdfdata/JAERI-Conf-2005-009.pdf. Accessed August 12, 2024

Uchikawa S (2007) Conceptual design of innovative water reactor for flexible fuel cycle (FLWR) and its recycle characteristics. J Nucl Sci Technol 44(3), 277–284. https://www.tandfonline.com/doi/pdf/10.1080/18811248.2007.9711283. Accessed August 12, 2024

Chapter 4
Light Water Reactor

4.1 Introduction

As discussed in Chap. 1, light water reactors, PWRs and BWRs, were developed and commercialized in the United States. This chapter describes light water reactors from the former Soviet Union/Russia, France, Japan, Germany, Sweden, South Korea, and China. US light water reactors were exported under license for the first units to France, Germany, Japan, and South Korea. Subsequently, these countries developed and commercialized their own light water reactors based on US light water reactors. Sweden developed its own BWR and imported PWR from the United States. While China developed small PWRs domestically, it also imported large PWRs from France, Russia and USA.

4.2 Light Water Reactor of the Former Soviet Union and Russia

In the former Soviet Union, a pressurized light water reactor called VVER was developed. They are used in the former Soviet Union and Eastern European countries, and in Finland. After becoming Russia, it has been built and operated in China, India, and Iran, and is under construction in Iran, Bangladesh, Egypt, Jordan, and Turkey. The VVER was initially researched at the Kurchatov Institute, and OKB Gidropress has been developing and manufacturing it. Since the first VVER started operating in Novovoronezh in 1964, 67 units have been built worldwide. In Russia, the Russian Federal Atomic Energy Agency (ROSATOM), as a state-owned company, is in charge of the construction of domestic and foreign nuclear power plants, uranium mining and enrichment, and nuclear fuel processing and supply, and Gidropress is a subsidiary of it. The first unit in Iran was originally a German pressurized light water reactor,

Y. Oka, *Nuclear Power Reactor Development*,
https://doi.org/10.1007/978-981-96-2731-8_4

but construction was interrupted in 1979 as a result of the Iran-Iraq war, and after many twists and turns, Russia took over the project and completed and operated it.

Many VVERs with an electrical output of 440 MW (VVER-440) and 1000 MW (VVER1000) have been built. The first VVER1000 started operating in 1980. Currently, VVER1200 and VVER1300, which are improved versions of VVER1000, are being built. The VVER is characterized by a horizontally placed steam generator, a hexagonal fuel assembly, no penetration holes such as measuring tubes at the bottom of the reactor pressure vessel, and a large capacity pressurizer. The V-230, an early plant of the VVER-440, had a containment volume that was insufficient to contain the steam released in the event of loss of coolant accident, so a steam condenser was added. VVER440/V-230 s in former East Germany and Bulgaria have been decommissioned after the collapse of the Soviet Union due to safety issues. Finland built two improved VVER440s called V-213, but after consideration, they built a vertical steam generator, their own containment, and a control system from the West.

The Tianwan Units 3 and 4 built in China are the first VVERs equipped with a core catcher that stores molten material falling from the bottom of the reactor pressure vessel in the event of a core meltdown. This core catcher is a structure filled with granules of iron and aluminum oxide between conical metals, and it cools and solidifies the core melt with its heat capacity. This ensures the containment performance of the reactor containment in the event of a core meltdown (ROSATOM 2015). This is a different method from the EPR's core catcher.

As of 2021 in Russia, there are 21 VVERs (3 VVER1200s, 13 VVER1000s, Five VVER440 reactors are in operation). In addition, there are 13 graphite-moderated, boiling light water-cooled pressure tube reactors (10 RBMK1000s and 3 EPG-6 cogeneration reactors with a thermal output of 62 MW), 2 sodium-cooled fast reactors (BN600 and BN800), and 2 floating cogeneration reactors called KLT40S with a thermal output of 35 MW. The KLT40S is a pressurized light water reactor based on the reactor of an icebreaker, supplying electricity and heating heat to naval bases in the Arctic Ocean. Nuclear power supplies 19% of Russia's electricity, and the power plants are operated by Rosenergoatom. In Russia, uranium mining, enrichment, fuel processing, decommissioning measures for nuclear power plants and nuclear submarines, and storage of spent fuel are also being carried out (IAEA 2023a).

4.3 French Light Water Reactors

In France, the French Electricity Authority (EDF) built six natural uranium fuel graphite-moderated carbon dioxide-cooled reactors developed under the leadership of the French Atomic Energy Commission (CEA) in the late 1960s, and used them for power generation in the early 1970s along with one pressurized light water reactor, one heavy water-moderated gas-cooled reactor, and one sodium-cooled fast breeder reactor (Phoenix). All six gas-cooled reactors had different designs, and their cost and operational issues were recognized by EDF's management. The pressurized light water reactor (Chooz-A, electrical output 310 MW) was built by a joint venture with

Belgium based on the license of Westinghouse's pressurized light water reactor and has been in operation since 1967. The first oil shock occurred in 1973. At that time, France was covering 75% of its electricity with thermal power using imported oil as fuel, but the price of oil quadrupled. This led to a government policy to cover electricity with nuclear power, and EDF built many large pressurized light water reactors, and France came to cover most of its electricity with nuclear power. The feature is standardization by building many reactors of the same design.

Along with the construction of many light water reactors, France built and operated a uranium enrichment plant called George Besse in 1978 to supply them with low enriched uranium fuel. This uranium enrichment method is a gas diffusion method, which requires a large amount of electricity, so a nuclear power plant was built on adjacent land, and about half of its generated electricity (800MWe) was used. The gas diffusion plant has been in operation for 33 years and was replaced in 2011 by the George Besse II uranium enrichment plant, which uses a centrifugal separation method that consumes less electricity. In Europe, URENCO, an international company, has been enriching uranium by centrifugal separation in the UK and the Netherlands since 1972 and in Germany since 1985. The French company (AREVA) acquired half of URENCO's shares in 2003, obtained uranium enrichment technology by centrifugal separation, and George Besse. II was constructed. France (ORANO, formerly AREVA's fuel cycle division) is not only involved in uranium enrichment and reprocessing, but also in the field of nuclear fuel production, supplying light water reactor fuel to the world as well as its own country.

France's light water reactors are, in chronological order, 900 MW class, 1300 MW class, and 1450 MW class, with 34, 20, and 4 units built respectively. Except for two 900 MW class nuclear power plants that were shut down in 2020 and were built near the border with Germany, all are still in operation. The 900 MW class is a 3-loop (3 steam generator systems) PWR, with designs called CP, CP1, and CP2 in chronological order, built from the 1970s to the early 1980s. CP0 and CP1 were built with a common control room and turbine building for two reactors. CP1 and CP2 have enhanced emergency cooling systems compared to CP0, and a total of 28 units have been built with the same technology.

These nuclear power plants are not turnkey contracts with the reactor manufacturer, but EDF is in charge of the entire construction, including civil engineering, by ordering the manufacture of the primary system of the reactor and the turbine generator from FRAMATOME and ALSTOM respectively, and conducting competitive bidding for the rest. It has been reported that cost reductions of 30–40% have been achieved through standardization of design and series manufacturing/construction. Standardization also has advantages in operation, such as reflecting operating experience, training operators, and inventory of replacement parts (Roche 2008). The 900 MW class PWR has been exported to South Africa, Korea, and China. China is developing and constructing a domestically produced pressurized water light water reactor called CPR-1000 based on this design and domestic construction and manufacturing experience.

Both the 1300 MW class PWR and the 1450 MW class PWR are 4-loop PWRs, called P4, P'4, and N4 respectively. N4 is a French design, not a Westinghouse license, and construction began from 1984 to 1991, it entered rated commercial operation from 2000 to 2002, and its electric output was increased to 1500 MW in 2003.

The EPR is a 1500 MW pressurized water light water reactor, which started operation in China in 2018 and in Finland in 2023 and 2024, and is under construction in France, and the UK. The EPR in France, Flamanville Unit 3 is connected to the grid in the summer of 2024. The EPR was initially developed in the 1990s jointly by a French company (AREVA) and a German company (Siemens-KWU). For example, the core catcher of the EPR (a device to spread and cool the molten material that flowed out from the bottom of the nuclear reactor pressure vessel in the event of a core meltdown) was developed by a German company. However, due to the change in Germany's nuclear power policy in 2000, the German company withdrew, and the EPR became an AREVA product. Subsequently, AREVA suffered a large loss due to the delay in the construction of the EPR in Finland and France, and now Framatome, which took over AREVA's reactor business (parent company is EDF), is in charge of design and manufacturing.

90% of EDF's power generation facilities are nuclear power. Therefore, in order to respond to the daily and nightly fluctuations in electricity demand and changes in electricity demand between weekdays and weekends, they perform load-following operations that vary the output of nuclear power plants. There are two methods of load-following operation: one using control rods and the other changing the boron concentration of the cooling water. The latter takes time. The former method has the problem that combustion (burn-up) at the bottom of the fuel assembly is difficult to progress, so they use control rods (gray control rods) with a small neutron absorption effect. With these methods, it is possible to change from 100% output to 30% output in 30 min. When the time for fuel replacement in the reactor core is imminent (at the end of the burnup cycle), the excess reactivity of the reactor decreases, so large output changes (load following) cannot be made. EDF combines a large number of light water reactors to perform load-following operations. Note that nuclear power plants and coal-fired power plants have a large proportion of capital costs, so the cheapest way to generate electricity is to operate at a constant output. Natural gas-fired and pumped-storage hydropower plants are suitable for load-following operation.

France covers about 70% of its electricity with nuclear power, and both generation costs and electricity rates are cheap. The first transmission lines between France, Switzerland, and Italy were created in the early twentieth century, but the European power network was strengthened with the liberalization of electricity in the late 1980s and the economic integration of European countries. Currently, electricity trading is taking place between Western European countries. French electricity is exported to Belgium, Germany, Switzerland, Italy, Spain, the UK, and others. On the other hand, because electricity is cheap in France, electric heating is widespread, and during extreme cold waves in winter, electricity demand surges, so there may be emergency imports of electricity.

4.4 Japanese Light Water Reactors

This section describes the overview of Japan's light water reactor. For information on Japan's nuclear research and development, energy use, computer-aided design in light water reactor design and construction, and the development of modular construction methods, refer to another book, "Nuclear Power and Society", Chap. 1.

The use of nuclear power generation in Japan began in 1963 when a boiling water type light water reactor power demonstration reactor (electric output 12.5 MW) imported from the United States started operating at the Japan Atomic Energy Research Institute. In 1966, the Japan Atomic Power Company's Tokai Power Station started commercial operation of a graphite-moderated carbon dioxide-cooled reactor (electric output 16.6 MW) imported from the UK, but Japanese power companies introduced American light water reactors, and in 1970, Kansai Electric Power's Mihama Unit 1 (PWR, electric output 340 MW) and Japan Atomic Power Company's Tsuruga Unit 1 (BWR, electric output 357 MW) were in operation, and in 1971, Tokyo Electric Power's Fukushima Daiichi Nuclear Power Station Unit 1 (BWR, electric output 460 MW) was in operation.

The light water reactor improvement standardization plan started around 1975, involving both the government and private sectors. The output of pressurized water reactors is classified into 550 MW (2 loops), 850 MW (3 loops), and 1100 MW (4 loops), and similar output classifications were made for boiling water reactors to increase their output. The third improvement standardization plan, which ended in 1986, developed the ABWR (electric output 1360 MW) and APWR (electric output 1500 MW). The ABWR began operation in 1996 as Unit 6 of the Tokyo Electric Power Company's Kashiwazaki-Kariwa Nuclear Power Plant, and has been in operation at Tokyo Electric Power Company, Chubu Electric Power, and Hokuriku Electric Power. The Shimane Unit 3 of Chugoku Electric Power and the Oma Unit 1 of J-Power are under construction, but they are required to meet the new regulatory standards established after the TEPCO Fukushima accident. The APWR began civil engineering work as Units 3 and 4 of the Japan Atomic Power Company's Tsuruga Power Station, but the plan has been suspended due to the accident at the TEPCO Fukushima Daiichi Nuclear Power Station in 2011. The manufacturer of PWRs is Mitsubishi Heavy Industries, Ltd., and the manufacturers of BWRs are Hitachi-GE Nuclear Energy, Ltd. and Toshiba Corporation. Mitsubishi Heavy Industries jointly developed the ATOMEA, a 3-loop PWR (electric output 1100 MW class), with a French company. A photo of the ABWR at Tokyo Electric Power Company's Kashiwazaki-Kariwa Nuclear Power Plant is shown in Fig. 4.1.

In Japan, 32 BWRs and 24 PWRs have been built and started operation as commercial nuclear power plants. However, after the TEPCO Fukushima accident in 2011, new regulatory standards were established, and it was decided that operation could be extended only once every 20 years. As a result, many power companies have decided to transition to decommissioning, mainly at smaller power plants. The number of light water reactors that have transitioned to decommissioning is 23, including three that were decommissioned before the accident. The remaining 33 are aiming to

Fig. 4.1 Advanced boiling water reactor (ABWR). TEPCO Kashiwazaki-Kariwa nuclear power plant

continue operation. In addition, there are three that were under construction when the TEPCO accident occurred. The number of power plants in operation that meet the new regulatory standards is increasing, but some power plants are still shut down for compliance with the new regulatory standards.

4.5 German Light Water Reactors

The development for the use of nuclear energy in West Germany began in the 1950s. Several nuclear research centers were established in Jülich, Karlsruhe, and other places from the late 1950s to the 1960s, and several types of research reactors and prototype reactors, including fast reactors, were built, and research on reprocessing and geological disposal of radioactive waste was also conducted. Research reactors were also installed at university research institutes.

In West Germany, Siemens developed PWRs under a license from Westinghouse (WH), and AEG developed BWRs under a license from General Electric (GE). The first experimental light water reactor for power generation was the VAK (electric output 16 MW) made by AEG/GE, which reached criticality in 1960. In 1961, domestic nuclear development began with Brown Boveri & Cie (BBC). Krupp installed a pebble-bed high-temperature gas reactor AVR (with an electrical output of 15 MW) at the Jülich Research Institute. Larger output high-temperature gas reactors were also built, but commercial operation was not successful, and light water reactors were used for nuclear power generation in Germany. The high-temperature gas reactor power plant that was built in China and started commercial operation in 2021 is based on the design and experience of Germany's pebble-bed high-temperature gas reactor.

In 1969, Siemens consolidated its nuclear business to create Kraftwerk Union (KWU) and developed domestic PWRs and BWRs. In 1968, a PWR with an electrical output of 340 MW started generating power, and in 1971 and 1972, BWRs and PWRs with an electrical output of 600 MW each started generating power. In 1981 and 1983, PWRs and BWRs with an output of 1200 MW each started commercial operation. German PWRs were built domestically and exported, operating in the Netherlands in 1973, Switzerland in 1979, and Spain in 1988.

The characteristics of German PWRs include the insertion of reactor instrumentation from the top of the reactor pressure vessel, the absence of penetrations for reactor instrumentation at the bottom of the reactor pressure vessel, and the adoption of a spherical steel reactor containment vessel, which is covered with a cylindrical reinforced concrete building. The developed 1300 MW PWR is called Konvoi, and three units were operating in 1988. Standardization is intended in Konvoi. Fourteen PWRs and ten BWRs were built in West Germany.

In East Germany, nuclear development began in the 1950s with the support of the former Soviet Union, and the first former Soviet PWR (with an electrical output of 62 MW) started operation in 1966. After that, five former Soviet PWRs (with an electrical output of about 400 MW) were built and operated from 1970 to 1976, but after the reunification of East and West Germany, they were abolished because they did not conform to the safety standards of the West. In total, 30 light water reactor power plants have been built in Germany. Before the TEPCO Fukushima accident in 2011, 17 nuclear power plants supplied about 25% of the electricity, but after that, a nuclear power phase-out policy was adopted, and all were planned to be shut down by the end of 2022. However, due to the impact of Russia's invasion of Ukraine, there were disruptions in the import of natural gas from Russia, so to ensure a stable supply of electricity, the shutdown deadline for some nuclear power plants was extended, but all were shut down in April 2023 (IAEA 2023b). In response to Russia's invasion to Ukraina, a policy to resume the use of nuclear power is being considered in 2025.

4.6 Swedish Light Water Reactors

The characteristic of Sweden's nuclear development is its independent development. This can be attributed to the fact that Sweden was an industrial nation, as seen in its steel and automobile industries. Sweden's nuclear program began in 1947 with a joint venture between the government and private companies; AB Atomenergi was established. In 1954, a research reactor was built at the Royal Institute of Technology in Stockholm, and in the 1960s, another was built in Studvik. The first nuclear power plant was the pressurized heavy water reactor Agesta, designed by the Swedish company ASEA. Agesta (with an electrical output of 12 MW and a heating output of 68 MW) was built in the bedrock underground of Stockholm and began commercial operation in 1964. It was mainly used for district heating, but due to the low price of oil at the time, a district heating plant using oil as fuel was built, and operation was stopped in 1973. However, immediately after that, the oil crisis occurred and the price of oil soared. The adoption of heavy water cooling was due to Sweden's possession of heavy water production technology and the advantage of being able to use natural uranium fuel. However, after the United States opened up uranium enrichment services to the world, the plan to build the Marvibicken nuclear power plant, which would have increased the output of the Agesta reactor tenfold, was abandoned in favor of using light water reactors for power generation. ASEA won its first BWR design contract in 1965. In 1968, the joint venture ASEA-Atom was established with the government, and its first nuclear power plant went into operation in 1972, with 12 units starting operation by 1985. Nine of these were BWRs designed by ASEA-Atom. The remaining three units were PWRs built under a license from Westinghouse. The choice of BWR as a domestic reactor was made because BWR does not require a steam generator and there is no problem with corrosion of its thin tubes. PWRs were also built because the state-owned power company that placed the order wanted to diversify its risks in terms of stable power supply.

In Sweden, the government decided to phase out nuclear power in 1988 due to the impact of the Chernobyl accident, but in 1991 and 1997, the government rescinded the phase-out for 2010 and simultaneously decided to shut down two nuclear power plants. The remaining nuclear power plants underwent construction to comply with new regulatory standards from 2000 to 2010. After the TEPCO Fukushima accident in 2011, nuclear power plants operating after 2020 were required to have an independent core cooling system. Subsequently, four nuclear power plants were shut down by 2020, citing reduced profits and the cost of regulatory compliance work. As of 2022, six nuclear power plants have obtained operating permits until 2040. New construction is also possible, but it has been decided to limit it to up to 10 units throughout Sweden (IAEA 2023c, Saido 1993, KSU 2000). Subsequently, in the fall of 2022, in response to the energy crisis caused by Russia's invasion of Ukraine, the government has been reviewing its energy policy, and a policy of expanding nuclear power generation has been decided. By 2045, it plans to double the power supply from nuclear power.

4.7 South Korea's Light Water Reactors

South Korea's use of nuclear power began in 1957 when it joined the IAEA. Since the 1970s, South Korea has been vigorously implementing nuclear power generation plans. First, they operated six 600 MW-class pressurized water reactors introduced from Westinghouse from 1978 to 1987, one CANDU reactor (electric output 600 MW class) introduced from Canada in 1982, three in the late 1990s, and two 900 MW-class pressurized water light water reactors introduced from France in the early 1980s. They

Fig. 4.2 Barakah nuclear power plant (UAE). *Source* Emirates nuclear energy corporation

have been operating 12 domestically produced 900 MW-class pressurized water light water reactors called OPR1000 since 1994, and two electric output 1400 MW-class pressurized water light water reactors called KPR1400 since 2016. The KPR1400 has been exported to the United Arab Emirates, where four units are in operation. Of the 26 units built in South Korea, two have ceased operation in 2017 and 2019 due to the influence of the anti-nuclear movement after the TEPCO Fukushima accident, and 24 units are in operation. Both OPR1000 and KPR1400 are based on the PWR of Combustion Engineering in the United States, and despite being large reactors, they are both two-loop (two steam generators) PWRs. Figure 4.2 shows the Barakah Nuclear Power Plant in the United Arab Emirates. The construction and operation permits were reviewed by experts with experience in the U.S. Nuclear Regulatory Commission and others at the UAE's Nuclear Regulatory Authority.

In South Korea, Korea Hydro & Nuclear Power Company (KHNP) is the only nuclear power company. The parent company is Korea Electric Power Corporation (KEPCO). In new construction, KHNP organizes the construction project, Korea Electric Power Corporation Engineering & Construction Company (KEPCO E&C) handles the design of the nuclear reactor steam supply system (NSSS) and plant construction, KEPCO Nuclear Fuel (KEPCO NF) handles the design and manufacture of nuclear fuel, Korea Atomic Energy Research Institute (KAERI) handles research and development, Doosan Heavy Industries & Construction (DHI) manufactures the NSSS, turbines, and generators, and Korea Institute of Nuclear Safety (KINS) supports safety regulations. The Nuclear Safety and Security Commission (NSSS), the nuclear safety regulatory commission, is established as an agency directly under the Prime Minister, independent of the promoting ministry (MOTIE, Ministry of Trade, Industry &Energy) (IAEA 2023d).

4.8 Chinese Light Water Reactors

China's nuclear power development began in the 1950s with the cooperation of the former Soviet Union. A nuclear research institute was established at the Chinese Academy of Sciences. After Soviet nuclear cooperation with China was cut off in 1959, China aimed for independent development in the early 1960s, and a nuclear power generation plan began in 1970. It stagnated during the Cultural Revolution that lasted from 1966 to 1977, but in 1991, the independently developed 300,000-kilowatt class PWR Qinshan Unit 1 and the 900,000-kilowatt class PWR Daya Bay Units 1 and 2 imported from France began commercial operation. Nuclear power generation expanded rapidly in the 11th Five-Year Plan from 2005. By 2022, 51 pressurized water reactors, two CANDU reactors, and one pebble bed high-temperature gas-cooled reactor are in commercial operation. Nineteen are under construction, and 33 are planned for construction, all of which are pressurized water reactors. In addition, at least 90 construction plans have been proposed. There are no decommissioned nuclear power reactors.

The main Chinese government agencies related to nuclear power generation are the China Atomic Energy Authority (CAEA), National Energy Authority (NEA), Ministry of Ecology and Environment (National Nuclear Safety Administration) (MEE/NNSA), CAEA, and National Health Commission. The CAEA oversees the nuclear industry and formulates nuclear policies and plans. The NEA is responsible for energy policy. The NNSA regulates nuclear power plants. The National Health Commission is responsible for preventing occupational radiation exposure and coordinating medical activities during nuclear emergencies. In 2013, the NEA was reorganized, a nuclear power generation department was created, and it oversees the entire nuclear power plan and promotes international cooperation.

State-owned enterprises involved in nuclear power generation are the China National Nuclear Corporation (CNNC), China General Nuclear Power Group (CGN), and State Power Investment Corporation (SPIC). CNNC and CGN own and operate most of the nuclear power plants. The State Nuclear Power Technology Corporation (SNPTC), which builds and operates the AP1000 under license from Westinghouse and is developing the domestically produced large-scale CAP1400, has been under SPIC since 2015.

The main research and development institutions for nuclear power generation are the China Institute of Atomic Energy (CIAE); Nuclear Power Institute of China (NPIC); Shanghai Nuclear Engineering Research and Design Institute (SNERDI); China Nuclear Power Technology Research Institute. The CIAE is located in Beijing, the NPIC is in Chengdu, both under CNNC, SNERDI is under SPIC, and the China Nuclear Power Technology Research Institute is a research and development institution under CGN.

The development and improvement of China's pressurized water reactors (PWRs) were carried out in parallel for small domestic reactors and large reactors imported from France. The former was developed by SNERDI and operated by CNNC in 1991, starting with the Taishan Unit 1 (the reactor pressure vessel was manufactured by a Japanese nuclear reactor manufacturing company), and the latter were the Daya Bay Units 1 and 2 imported from France. Taishan Unit 1 is a pressurized water reactor with an electrical output of 300 MW, and four units have been exported to Pakistan. Subsequently, a 600 MW class was developed, and four units were built in the second phase of the Taishan project and two units on Hainan Island. In the third phase of Taishan, two Canadian heavy water reactors called CANDU6 were built and commercial operation started in 2002 and 2003.

The Daya Bay Units 1 and 2, which started operation in 1993, are French 900 MW class PWRs, known as the M310 type. Subsequently, two units started operation in 2002 and 2003 at the Lingao Nuclear Power Plant adjacent to the Daya Bay Nuclear Power Plant. The second phase of the Lingao Nuclear Power Plant involved the construction of four CPR1000 units, which have a higher domestic production ratio, in collaboration with French companies, and operation started in 2010. Fourteen CPR1000 units have been built and are in operation. All of these are nuclear power plants of CGN. Eight improved M310 power plants from France have also been built and are in operation by CNNC. The CPR1000 has evolved into the improved CPR1000 and ACPR1000, with two and four units built and operated respectively, but based on these, CGN and CNNC have cooperated to work on exports, developing the Hualong One (also known as HPR1000). The first two units were built in China and started operation in 2021, and two more units have been exported to Pakistan and have been in operation since 2021. In addition, twelve 1200 MW Hualong 1 units are currently under construction throughout China.

In China, so-called third-generation light water reactors are also being built. This began in 2004 when the State Council conducted an international competitive bidding for the type of reactor to be built at the Sanmen and Yangjiang nuclear power plants. As a result, the AP1000 pressurized water reactor from Westinghouse (WH) in the United States and the EPR pressurized water reactor from AREVA (at the time) in France were built. Some of the equipment for these reactors, such as reactor pressure vessels and turbine systems, were supplied by manufacturers in Korea and Japan.

The AP1000 is a pressurized water reactor that has enhanced passive safety, as described later. It adopted canned motor pumps for the primary cooling water pumps for the first time for nuclear power plants. The canned pump does not require cooling water for the pump shaft seal and eliminates the loss of coolant accident from the pump seal. It, however, struggled to meet the safety requirement for flow rate reduction at flow coast down accident. The construction of AP1000 that began in 2008 was delayed, but it has been in operation since 2018. Currently, four units are in operation. Technology transfer from WH to SNTPC has also been carried out, and the construction of CAP1400, which has increased the output to 1400 MW, is underway. The construction of the EPR (electric output 1660 MW) was carried out in a joint venture between a company under CGN and AREVA. Although the start of construction was later than the European EPR, it was the first EPR to operate in the world in 2018, and two units are in operation. Two units of Russia's VVER1000 have been in operation since 2006, two units since 2017, and four units of VVER1200 are under construction.

China's nuclear power plants have been built along the coast. The nuclear power plants planned in the inland areas have been postponed since the TEPCO Fukushima accident in 2011, but they have not been canceled. Except for the ones built in the early stages, all of China's nuclear power plants are large light water reactors, and they have been built in sets of two. In the case of small pressurized water reactors with steam generators inside the reactor pressure vessel, the construction of one ACP100 (electric output 125 MW) unit began in 2019 by a venture company under CNNC. The location is a site where a domestically produced 600 MW class PWR is being built and operated on Hainan Island. The nuclear fuel assemblies for China's nuclear power plants can be produced and supplied domestically (IAEA 2022a, WNA 2024a).

References

IAEA (2022a) Country nuclear power profiles: China, IAEA. https://cnpp.iaea.org/public/countries/CN/profile/preview. Accessed August 12, 2024

IAEA (2023a) Country nuclear power profiles: Russian Federation, IAEA. https://cnpp.iaea.org/public/countries/RU/profile/preview. Accessed August 12, 2024

IAEA (2023b) Country nuclear power profiles: Germany, IAEA, 2023. https://cnpp.iaea.org/public/countries/DE/profile/preview. Accessed August 12, 2024.

IAEA (2023c) Country nuclear power profiles: Sweden, IAEA. https://cnpp.iaea.org/public/countries/SE/profile/preview. Accessed August 12, 2024

IAEA (2023d) Country nuclear power profiles: Republic of Korea, IAEA. https://cnpp.iaea.org/public/countries/KR/profile/preview. Accessed August 12, 2024

KSU (2000) Operating experience from swedish nuclear power plants, Karnkraftsakerhet and Education AB (KSU) 2000. https://inis.iaea.org/collection/NCLCollectionStore/_Public/32/018/32018396.pdf. Accessed August 12, 2024

Roche B (2008) The French nuclear program: EDF'S experience. http://apw.ee.pw.edu.pl/tresc/-eng/08-FrenchNucleamProgram.pdf. Accessed August 12, 2024

ROSATOM (2015) The VVER today: evolution, design, safety. ROSATOM overseas. https://www.rosatom.ru/upload/iblock/0be/0be1220af25741375138ecd1afb18743.pdf. Accessed August 12, 2024

Saido K, Gray JE (1993) The miracle of nuclear power, Part 5 in Sweden, Nikkan Kogyo Shimbun, 1993 (in Japanese)

WNA (2024a) Nuclear power in China, world nuclear association, February 2024. https://world-nuclear.org/information-library/country-profiles/countries-a-f/china-nuclear-power.aspx. Accessed August 12, 2024

Chapter 5
Passive Safety Light Water Reactors and Severe Accident Countermeasures

Light water reactors with passive safety have been studied in large numbers since the early 1980s. Passive safety is "the function of ending an event by a mechanism that uses natural phenomena occurring in the reactor when an accident occurs in the reactor, without depending on external power or operator action". Severe accident is a type of accident involving the reactor core degradation and meltdown in nuclear power plants when radioactive release to the public will be notable. In response to the Chernobyl accident, severe accident countermeasures for light water reactors were considered. This chapter describes some of them introduced in modern LWRs.

In light water reactors, an emergency cooling system is provided to cool the reactor core in the event of an accident, and its pump is activated to ensure the cooling of the nuclear reactor. An emergency power source (diesel generator) for driving the pump is also prepared. These devices are dynamic devices with rotating parts and so on. The valves installed in the piping of the emergency cooling system are also dynamic devices. This valve also needs to be opened when the emergency pump is started. Large nuclear power plants such as light water reactors ensure safety during accidents with dynamic devices. Ensuring safety with dynamic devices is called dynamic safety. To ensure dynamic safety, nuclear power plants require pumps, valves, and emergency power sources that are not necessary for thermal power generation. These safety system facilities need to be made to meet the technical standards for them, and the burden of maintenance management to maintain their functions is also large. For example, it is said that nuclear power plants require 10 times as many valves as thermal power plants of the same output scale.

Regardless of whether it is for research or power generation, a nuclear reactor is designed to have inherent safety (negative feedback characteristics) that automatically suppresses output increase without operating control rods as its output increases. With this characteristic, for example, if the cooling of the nuclear reactor is lost, the nuclear fission chain reaction is suppressed. The control rods stop the nuclear reactor when accidents are detected. However, even after the nuclear fission chain reaction is stopped, the nuclear fuel heats up a little due to the radioactive decay of the

Y. Oka, *Nuclear Power Reactor Development*,
https://doi.org/10.1007/978-981-96-2731-8_5

fission products. To remove this, it is necessary to open the valve of the emergency core cooling system piping and start the pump to cool the reactor core. For redundancy, multiple dynamic safety systems such as emergency core cooling systems are provided in nuclear power plants. However, as in the case of the TEPCO Fukushima accident, if all the power sources for driving the pump are lost, the decay heat cannot be removed, and the nuclear fuel rods melt. Water is chemically reduced by the metal of the fuel cladding tube that has become high temperature, and hydrogen and oxygen are generated. The pressure inside the nuclear reactor containment vessel rises, and volatile fission products leak. The radioactive substances of a nuclear power plant are the fission products in the nuclear fuel, and it is required to prevent their releasing into the environment during an accident.

Dynamic safety systems such as emergency cooling systems that are provided to prevent the releasing of large amounts of radioactive substances into the environment in the event of an accident require a power source for driving the pump. PIUS, a passive safety light water reactor concept that removes decay heat using natural forces without using a dynamic safety system was announced by a Swedish nuclear reactor manufacturer in the early 1980s. In the late 1980s, nuclear reactor manufacturers and research and development institutions in various countries, such as WH (Westinghouse) and GE (General Electric) in the United States and Siemens KWU in Germany, conducted research and development on passive safety light water reactors and passive safety devices. Among these, PIUS, AP600, and AP1000 developed by WH, SBWR and ESBWR developed by GE, SWR100 (KERENA) considered by Siemens KWU (later German Areva), and severe accident countermeasure facilities are described in the following. Note that natural forces refer to differences in fluid pressure, gravity, etc.

PIUS is a pressurized water light water reactor using a large-capacity nuclear reactor pressure vessel made of prestressed concrete, which can remove decay heat for a long period of time due to the large amount of water stored in the nuclear reactor pressure vessel. While the nuclear reactor pressure vessels of regular light water reactors are made of steel, by covering the concrete walls with many steel wires and tightening the steel wires, compressive stress is applied to the concrete, making it possible to withstand high internal pressure. This allows for the creation of a large-capacity vessel using prestressed concrete for the nuclear reactor pressure vessel. A prestressed concrete vessel has been developed in the nuclear field as a containment vessel which contains a nuclear reactor pressure vessel and steam generators, etc. and in Japan, it is used as the containment vessel for 1200 MW-class pressurized water light water reactors. Prestressed concrete is used in fields other than nuclear power, such as for bridges.

The passive safety operating principle of PIUS is shown in Fig. 5.1. During normal operation, the nuclear reactor primary coolant and the water pool on the outside of the cylinder surrounding the core are balanced by the pressure difference, and the boundary between the hot primary coolant and the cold pool water is formed within the lower density lock. In the event of an accident, this balance is broken, pool water flows in from the density lock, and the core is cooled. This allows the nuclear reactor to be shut down and decay heat to be removed even without pumps, valves, or their

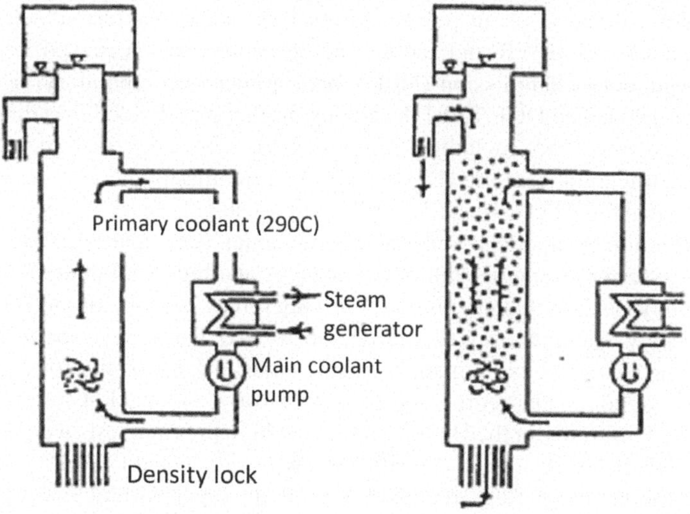

Fig. 5.1 Passive safety principle of PIUS reactor. *Source* ATOMICA, JAEA (in Japanese)

power sources. The pool water contains boron, and the neutron absorption effect of boron can stop the nuclear reactor. Reactivity control of the nuclear reactor is done not by control rods, but by adjusting the boron concentration in the primary coolant and the coolant temperature. The diameter of the prestressed concrete nuclear reactor pressure vessel is 12 m, the height is 43 m, the wall thickness is 7 m, and the operating pressure is 9 MPa, which is lower than that of a regular pressurized water light water reactor (15.4 MPa). The steam generators (straight tube type, 4 units) are installed outside the nuclear reactor pressure vessel. The primary coolant pump is a recirculation pump used in Sweden's BWR. The electrical output is 640 MW (IAEA 1994a; Pedersen 1989). Efforts were made to acquire customers for the construction of PIUS, but it did not lead to construction. However, it triggered the development of passive safety reactors and research on passive safety technology. Research and development of PIUS began before the Chernobyl accident in 1986. Equipment used in safety systems requires high reliability, so the quality control requirements during manufacturing are strict and expensive. Passive safety reactors aim not only for safety but also for cost reduction and simplification of maintenance and operation management by eliminating dynamic equipment from safety systems.

Advanced Passive pressurized water (AP600): The development of the reactor began in 1985 by Westinghouse (WH). The development was carried out within the Advanced Light Water Reactor (ALWR) program of the U.S. Department of Energy (DOE) and the Electric Power Research Institute (EPRI). EPRI compiled the design requirements of power companies for new light water reactors from the 1990s onwards into an ALWR Utility Requirement Document. The first volume summarizes the important design requirements, the second volume is about conventional large light water reactors (electric output 1350 MW) with dynamic safety systems, and the third volume is about passive safety light water reactors (electric output 600 MW), discussing PWR and BWR. This document was reviewed by the U.S. Nuclear Regulatory Commission (NRC), and opinions on regulatory issues were published in 1993. The DOE/EPRI development plan was divided into three phases over 11 years from 1985, including design development, confirmation testing, and obtaining certification from the U.S. Regulatory Commission.

The design of the AP600 was led by WH, with a team of plant engineering companies, construction companies, and power companies participating. Passive safety systems have fewer problems with power outages and failures, but are weaker than dynamic safety systems that use pumps as driving forces. The operation confirmation of passive safety equipment was first carried out by WH, but the tests for design assurance were carried out by participating organizations internationally for each theme. The test of the emergency core cooling system was conducted at low pressure at Oregon State University in the United States, and high pressure tests were conducted at SIET in Italy. Furthermore, systematic tests were conducted from 1992 to 1999 at the Japan Atomic Energy Research Institute (now the Japan Atomic Energy Agency) using an existing thermal fluid experiment facility called the Large Scale Test Facility (LSTF) with AP600 equipment added. Experiments on the containment of core melt in the reactor pressure vessel, an event outside the design basis, were conducted at the University of California, Santa Barbara, commissioned by the NRC. SIET is a large thermal fluid experiment facility located within a thermal power plant, capable of conducting large-scale experiments with a large supply of steam and purified water from the thermal power plant.

The main features of the AP600 are the use of canned motor pumps for the primary cooling water pump and the adoption of passive safety systems. Conventional pumps required a pump shaft seal and a seal water system, but with the adoption of canned motor pumps, these became unnecessary, simplifying the system and making maintenance easier, and there was no need to consider cooling water leakage accidents from the seal part. The safety system is composed of a passive safety system that uses compressed gas pressure, gravity, and natural circulation force. Two core water supply tanks (small-capacity, high-pressure injection system driven by gravity), two pressurized tanks (medium capacity, medium pressure injection system driven by compressed gas), and a refueling water storage tank inside the containment vessel (large-capacity, low-pressure injection system driven by gravity) are provided. These operate in the event of a loss of cooling water accident.

Figure 5.2 shows the passive containment cooling system of AP600. In the long term after an accident, the reactor pressure vessel and primary cooling system are flooded up to the top by the water in the refueling water storage tank, allowing the reactor core to be cooled. In the event of an accident other than loss of cooling water, heat removal is first performed by the steam generator, and if this is not possible, heat is removed by natural circulation through the residual heat removal heat exchanger installed in the containment refueling water storage tank. The heat released into the containment in this way is released into the atmosphere by air naturally circulating through the natural ventilation path provided between the containment wall and the building outside it. A passive containment cooling water storage tank is installed

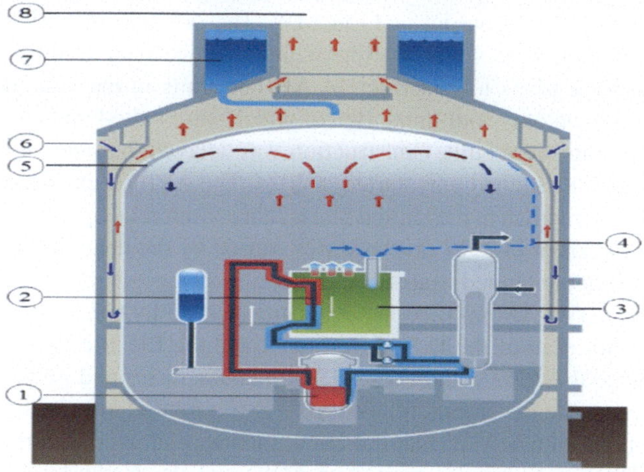

Fig. 5.2 Passive containment colling system of AP1000. *Source* Vereb et al. (2012) The AP1000 nuclear power plant innovative features for extended station blackout mitigation, proceedings of ICAPP' 12, Chicago USA, June 24–28, 2012, p. 12,063

at the top of the containment building, which promotes heat removal by dropping water droplets onto the outer wall of the containment by gravity. The AP600 also has dynamic devices such as cooling water pumps, but they are not safety systems, so construction costs and maintenance work are reduced. In the event of an accident, these dynamic devices first operate to cool the nuclear fuel. There is a focus on passive safety, but simplification of equipment and maintenance is intended (IAEA 1994b).

The AP600 received final design certification from the NRC in 1998. However, there were no orders from power companies. At that time, many small capital cost gas turbine combined cycle power plants were being built. Gas turbine combined cycle power plants are a type of power plant that enlarges gas turbines developed as aircraft engines for power generation, leads the waste heat of the gas turbine to a waste heat recovery boiler to make steam, and drives a steam turbine, using both a gas turbine and a steam turbine. Since gas turbines have developed as aircraft engines, startup, shutdown, and maintenance are easy. Gas turbine combined cycle power generation has a high proportion of fuel costs in power generation costs and a high power production cost (power generation cost excluding construction cost amortization), but the construction cost (investment risk) is small, so it is suitable for the investment environment of power liberalization and economic liberalization, and many were built in the United States. Many were also built in Europe and Japan afterward. Gas turbine combined cycle power generation is an innovation in power generation technology.

In the United States, in 2002, the Nuclear Power 2010 program was launched with the aim of starting operation of new nuclear power plants by 2010. The NP2000 (2010) plan was initiated. In response to this, WH designed the AP1000, a larger version of the AP600. The electrical output is 1154 MW. The NP2000 plan also included the construction of large light water reactors with dynamic safety systems. As of around 2015, there were plans to construct 14 AP1000 units in the United States, but construction began on only four units, two of which were halted midway through construction. Two units were built. Construction was delayed, but one unit started operation in 2023, and the second unit started in 2024. The AP1000 exported to China started construction later than in the United States, but commercial operation began in 2018. WH went bankrupt in 2017 due to losses caused by the delay in the construction of the AP1000 in the United States.

The development of the SBWR (Simplified BWR) began in 1985 by GE, as part of the new light water reactor plan of the U.S. government and the Electric Power Research Institute, similar to the AP600. Japanese BWR manufacturers and power companies also cooperated in the development. The SBWR is a natural circulation BWR with an electrical output of 670 MW. Natural circulation was adopted in early BWRs such as Japan's JPDR (electrical output 12.5 MW) and the Netherlands' DODEWAARD (electrical output 58 MW). The electrical output of the SBWR is 600 MW. Compared to forced circulation BWRs, in order to secure the driving force of the cooling water by natural circulation, the effective length of the core fuel is shortened to reduce pressure loss, and a long chimney section is provided at the top of the core. As a result, the reactor pressure vessel has become longer. As a result, the reactor pressure vessel has become larger and its water holding capacity has increased.

The SBWR adopts a passive safety system. First, in the event of a reactor isolation event (such as when the main steam isolation valve is quickly closed to prevent turbine damage due to overspeed when power transmission is not possible due to a failure in the power transmission system from the nuclear power plant and there is no steam turbine load), the reactor is shut down with control rods, the main steam isolation valve is closed, and the main steam is released and condensed in the pressure suppression pool of the containment vessel. Then, using an isolation condenser (IC), the steam generated from the core by residual heat is led to a condenser placed in a water pool on the outside upper part of the containment vessel, condensed, and the condensed water is returned to the reactor pressure vessel by its own weight. The heat transferred to the water pool is radiated into the atmosphere. The pressure rise at the time of closing the main steam isolation valve is gradual because the reactor pressure vessel is large. The isolation condenser was used in early BWRs such as the Fukushima Daiichi Nuclear Power Station Unit 1 of TEPCO.

In the event of a loss of coolant accident, the depressurization valve is operated by a water-level drop signal in the reactor pressure vessel, and after depressurization, the water in the GDCS pool provided in the containment vessel is supplied to the reactor pressure vessel by gravity using the gravity-driven cooling system (GDCS). This piping system is normally closed, so the flow path is opened by a rupture valve. Long-term decay heat removal after an accident is the passive containment cooling system (PCCS). The system uses PCCS. PCCS condenses the steam generated in the containment in a heat exchanger installed in a water pool outside the containment, and puts the condensed water into the GDCS pool inside the containment. In addition

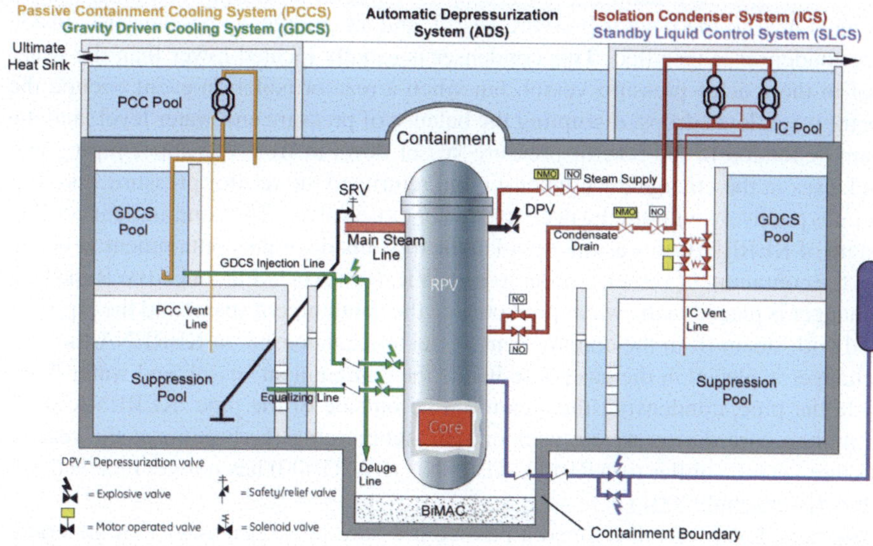

Fig. 5.3 Passive safety system of ESBWR. *Source* IAEA, status report 100—economic simplified boiling water reactor (ESBWR) 2011

to these passive safety systems, the high-pressure water injection function of the control rod drive system installed as a regular system, and the low-pressure core water injection function of the fuel and auxiliary pool cooling system are used, so that for the period when AC power can be used, the impact of accidents can be reduced by operator operation, and it is designed to be able to avoid activating the passive emergency core cooling system during many abnormal transient changes (IAEA 1994c).

SBWR also did not get customers and did not lead to construction. In the 2000s, GE announced the Economic Simplified Boiling Water Reactor (ESBWR), which is a larger version of the SBWR. The electric output is 1530 MW. The passive safety system of ESBWR is shown in Fig. 5.3. Although not described in the figure, the control rods are inserted from the bottom of the reactor pressure vessel as in a normal BWR. The position of the pressure suppression pool of the containment vessel is slightly higher than that of a normal BWR or ABWR (GE Hitachi 2011). ESBWR obtained NRC design certification in 2014. Some power companies have obtained NRC early site permits for ESBWR construction, but construction has not started. GE/HITACHI Nuclear Energy Corporation has designed the small light water reactor BWRX-300 based on the experience of these reactors, and has been working to obtain NRC approval since 2020.

Research and development of passive safety equipment was also conducted at Siemens KWU in Germany (later AREVA in Germany). Figure 5.4 shows the containment vessel and passive safety system of KERENA (SWR1000). KERENA is a forced circulation BWR with an electrical output of 1000 MW, developed from 1995, incorporating a passive safety system. In KERENA, the core flooding pool is arranged next to the reactor pressure vessel, and an emergency condenser is inside it. A condenser is installed. This condenser is usually located lower than the water level in the reactor pressure vessel, but when a reactor isolation event occurs, the reactor water level drops, disrupting the balance of pressure and water level, and the steam generated in the reactor pressure vessel flows in from the upper piping and condenses in the emergency condenser and returns to the reactor pressure vessel. It operates passively without the need to open or close valves. The containment cooling system of KERENA uses a water pool at the top outside of the containment vessel to cool the containment vessel, similar to ESBWR. However, while in ESBWR the heat exchanger is placed in the water pool outside the containment vessel and the piping is filled with steam from the containment vessel and condensed, in KERENA the heat exchanger is placed in the gas phase inside the containment vessel, and water flows inside the pipe, condensing the steam on the outside of the pipe. KERENA cools the molten core during severe accidents by submerging the bottom of the reactor pressure vessel, similar to AP1000. KERENA (SWR1000) has not yet received any orders (Brettschuh 2001).

Siemens KWU has developed a passive pressure pulse generator that generates pressure pulse signals using changes in the water level inside the BWR reactor pressure vessel. This device can generate signals without the need for a power source for the measuring instrument. In Japan, a researcher at Mitsubishi Heavy Industries has devised a new accumulator that passively switches the flow rate of the accumulator used in pressurized water reactors using the resistance of vortices created by water pressure.

We have been discussing passive safety light water reactors, but this is because they have been widely researched worldwide since the Chernobyl accident, and the author is not saying that passive safety light water reactors are the future direction of light water reactors. Passive safety devices and methods are also used in existing light water reactors, such as accumulators of pressurized water reactors and gravity-driven control rod insertion mechanisms. There are various classifications for the definition of passive safety, but for example, valves are widely used not only in existing light water reactors but also in passive safety light water reactors. Many valves are needed in power reactors to realize the operation logic of safety systems. Valves have movable parts and cannot be called completely passive devices.

Passive safety light water reactors, as seen in the examples of AP1000 and ESBWR, have large water tanks on top of the reactor building. This affects the seismic resistance requirements of the building. In earthquake-prone countries, earthquake assumptions are highlighted in the approval process. There is still uncertainty about the intensity and frequency of possible earthquakes. The use of buildings with vibration isolation structures requires regulatory approval and has not yet been applied to reactor buildings. Passive safety is also used in conventional light water reactors (dynamic safety light water reactors), such as pressurizers and gravity-driven (drop) control rods. Not only passive safety light water reactors, but also existing light water reactors are future options for light water reactors. In many countries around the world, construction of dynamic safety light water reactors is underway. This is not to say that passive safety light water reactors are bad.

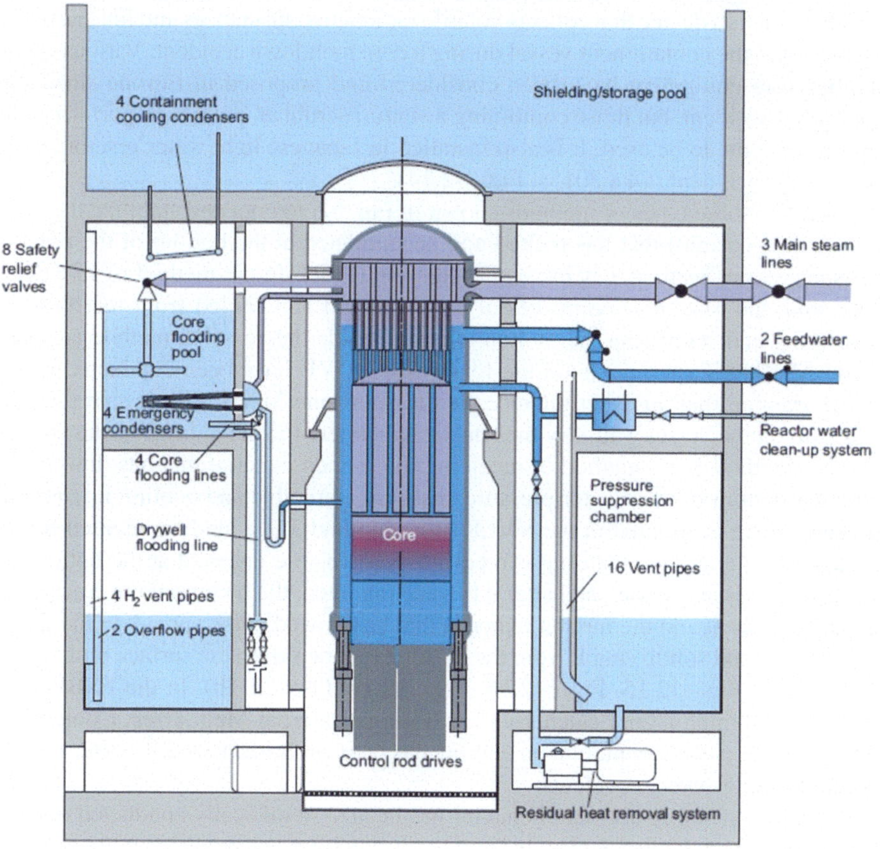

Fig. 5.4 Passive safety system of KERENA. *Source* Zacharias et al. (2012) KERENA safety concept in the context of the Fukushima accident. In: Proceedings of ICAPP'12, Chicago USA, June 24–28, 2012, p. 12,366

Since the Chernobyl accident, research on core meltdown (severe accident) behavior has been conducted in Europe and other countries, and equipment to reduce the impact of severe accidents has been developed by light water reactor manufacturers in various countries, including Siemens KWU. Some of them are introduced in the following.

The catalytic hydrogen recombiner is a device that uses a catalytic reaction to return the hydrogen released into the containment vessel during a severe accident to water, and is used to reduce the hydrogen concentration in the containment vessel to prevent explosive combustion of hydrogen. Conventional electric hydrogen recombiners which need power supply have been used, but since it is catalytic, no power source is needed (Oka 2015a, Fig. 12.13). In Japan, after the TEPCO Fukushima accident, it was added to existing light water reactors in response to new regulatory standards, and is used together with electric ones.

Filter vent is a device that collects volatile radioactive substances and fine particles released into the containment vessel during a core meltdown accident. Various types of filter vent equipment have been considered and proposed in Europe since the Chernobyl accident, but those combining a venturi scrubber and a high-performance filter have come to be used. It is also installed in Japanese light water reactors after the TEPCO accident (Oka 2015a, Fig. 12.14).

There are three types of methods shown in Fig. 5.5 to cool and stabilize the core melt (called corium) that has melted and accumulated at the bottom of the reactor pressure vessel. Method 1 is in-vessel retention (IVR): In the method of In-Vessel Retention, the reactor pressure vessel is submerged and cooled from the outside, solidifying and stabilizing the molten material inside the vessel, a method adopted in the AP1000. Experiments on the feasibility of IVR have been conducted using actual uranium melt in experiments called Rasplav and MASCA. It is said that the heat load on the vessel wall becomes maximum where the metal layer of the melt is formed. Method 2 is a method of receiving the molten material that has flowed out from the bottom of the reactor pressure vessel in a crucible and cooling it from the outside, which is adopted in the VVER1200. Method 3 is a method adopted in the EPR, where the molten material is received in a crucible installed at the bottom of the reactor pressure vessel, and after it has accumulated, the bottom of the crucible is damaged by heat and the molten material that has flowed out is spread on the floor, first cooling and solidifying it from the bottom and then from the surface with water (Oka 2015a, Fig. 12.15, Fig. 12.16, Fig. 12.17, Oka 2015b). In the ESBWR, a method combining a core catcher called Basemat-internal Melt Arrest Coolability (BiMAC) and passive containment cooling has been proposed to stabilize the molten material during severe accidents.

Research on severe accident behavior has been systematically conducted mainly in Europe and the United States. In Europe, research on severe accident phenomena has been conducted mainly by the Cadarache Research Institute of the French Atomic Energy Commission and the Karlsruhe Institute of Technology in Germany (formerly the Karlsruhe Research Center). A research network called SARNET has been established and is active in the EU research framework (Miassoedov 2012; IRSN 2007). English books have also been published (Sehgal 2012). Severe accident research in the United States has been conducted with the support of the NRC, DOE, and the nuclear industry at Sandia National Laboratories, Argonne National Laboratory, and EPRI, with a view to reflecting it in severe accident calculation codes such as MELCOR and MAAP (Farmer 2016, Gauntt 2019). Severe accident calculation codes like MELCOR can be considered as a way to accumulate and utilize knowledge of severe accident phenomena.

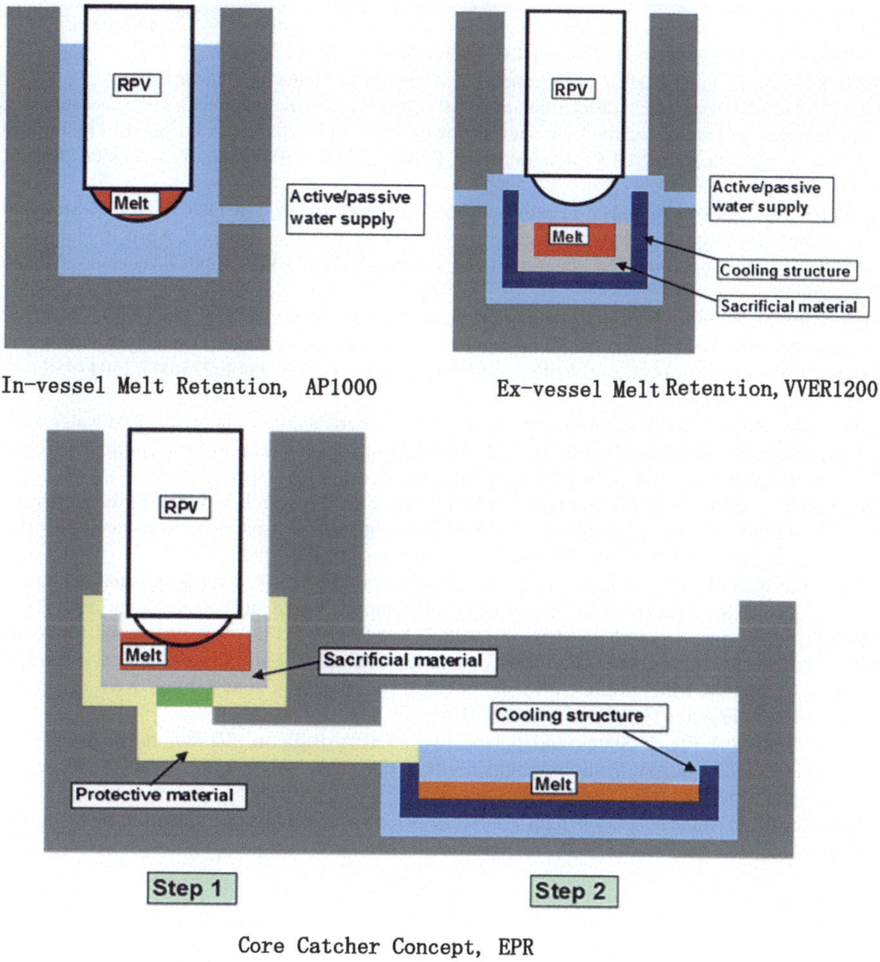

Fig. 5.5 Three types of melt stabilization measures. *Source* Oka and Bittermann, "Implications and lessons for advanced reactor design and operation", of "Reflections on the Fukushima Daiich nuclear accident", Springer open, 2014, Chap. 12

References

Brettschuh W (2001) The SWR 1000: a nuclear power plant concept with boiling water reactor for maximum safety and economy of operation. https://www.osti.gov/etdeweb/servlets/purl/20235349. Accessed August 12, 2024

Farmer MT, Corradini M, Gauntt RO (2016) US Department of Energy Severe Accident Research in the areas of in-vessel core melt progression and ex-vessel debris coolability, IAEA Technical Meeting on Phenomenology and Technologies Relevant to In-Vessel Melt Retention and Ex-Vessel Corium Cooling, SNERDI, Shanghai, China. https://inis.iaea.org/collection/NCLCollectionStore/_Public/51/089/51089941.pdf?r=1. Accessed August 12, 2024

Gauntt RO (2019) Severe accident modeling and evolution, a historical survey, SAND2019–4551PE. https://www.osti.gov/servlets/purl/1648750. Accessed August 12, 2024

GE Hitachi (2011) The ESBWR plant general description. GE Hitachi Nuclear Energy.

IAEA (1994a) PIUS technical description and cost estimates, IAEA-TECDOC-739, Case study on the feasibility of small and medium nuclear power plants in Egypt Annex G, April 1994. https://inis.iaea.org/collection/NCLCollectionStore/_Public/25/047/25047958.pdf. Accessed August 12, 2024

IAEA (1994b) AP600 technical description and cost estimates, IAEA-TECDOC-739, Case study on the feasibility of small and medium nuclear power plants in Egypt, Annex E, April 1994. https://inis.iaea.org/collection/NCLCollectionStore/_Public/25/047/25047958.pdf. Accessed August 12, 2024

IAEA (1994c) SBWR, Technical description and cost estimates, IAEA-TECDOC-739, Case study on the feasibility of small and medium nuclear power plants in Egypt, Annex F, April 1994. https://inis.iaea.org/collection/NCLCollectionStore/_Public/25/047/25047958.pdf. Accessed August 12, 2024

IRSN (2007) Research and development with regard to severe accidents in pressurised water reactors: Summary and outlook, Report IRSN-2007/83. https://inis.iaea.org/collection/NCLCollectionStore/_Public/43/012/43012266.pdf?r=1&r=1. Accessed August 12, 2024

Miassoedov A, et al (2012) Overview of LWR severe accident research activities at the Karlsruhe institute of technology, proceedings of ENCIT 2012. https://inis.iaea.org/collection/NCLCollectionStore/_Public/47/056/47056916.pdf. Accessed August 12, 2024

Oka Y, Bittermann D (2015a) Implications and lessons for advanced reactor design and operation. In: J Ahn (ed) Reflections on the Fukushima Daiichi nuclear accident chapter 12, Springer, 2015. https://link.springer.com/chapter/10.1007/978-3-319-12090-4_12. Accessed August 12, 2024

Oka Y (2015b) Perspectives on the safety of nuclear facilities, "Earthquake Resilience Nuclear Facilities" Pacific Rim Forum, University of California Berkeley, June 8. http://www.aec.go.jp/jicst/NC/about/kettei/150608.pdf. Accessed August 12, 2024

Pedersen T (1989) PIUS: Status and perspectives, IAEA Bulletin, 3/1989. https://www.iaea.org/sites/default/files/publications/magazines/bulletin/bull31-3/31302392529.pdf. Accessed August 12, 2024

Sehgal BR (2012) Nuclear safety in light water reactors: severe accident phenomenology, Academic Press 2012

Chapter 6
Heavy Water-Moderated Reactors: CANDU, Etc.

This chapter mainly discusses the CANDU, a power reactor developed by Canada that uses heavy water for moderation and cooling, and also introduces other power reactors that use heavy water for moderation. The features of heavy water-moderated reactors are that they can use natural uranium fuel or low enriched uranium fuel, and that they use pressure tubes instead of a reactor pressure vessel. The problem of positive reactivity coefficient in heavy water-moderated light water-cooled reactors and its countermeasures is also discussed.

Heavy water has a lower neutron absorption rate than light water, allowing natural uranium to be used as reactor fuel. Canada has developed a pressurized heavy water reactor (PHWR), also known as CANDU, which uses heavy water for both moderation and cooling. In 1955, they operated the first experimental power reactor (electric output 22MW). In 1967, they operated one prototype reactor of the CANDU type with an electric output of 250MW, and in 1972, they operated one prototype reactor of the heavy water-moderated boiling light water-cooled type (Gentilley Unit 1, SGHWR type, generated electricity and decommissioned in 1972 and 1977). After that, they operated CANDU type power plants of 600MW class and 800MW class, and as of 2022, 19 units are in operation, covering about 15% of Canada's electricity demand, with most of the nuclear power plants located in Quebec. Many power plants started operation from the late 1970s to the early 1990s. After that, many are still in operation after major repairs. A total of 12 CANDUs have been exported to overseas, including South Korea, Romania, India, Pakistan, Argentina, and China.

CANDU is a heavy water-moderated and cooled pressure tube type, and its configuration is shown in Fig. 6.1. The fuel assembly is a bundle of fuel rods arranged in a circle (50 cm long, 10 cm in diameter), which is inserted sideways into a zirconium metal tube called a pressure tube. Inside the pressure tube, the fuel rods are cooled by heavy water under high temperature and pressure. On the outside, there is a horizontally placed cylindrical tank called a calandria that contains heavy water as a moderator. Many pressure tubes are arranged at equal intervals inside the calandria. Around the pressure tube, calandria tubes filled with carbon dioxide are arranged,

Y. Oka, *Nuclear Power Reactor Development*,
https://doi.org/10.1007/978-981-96-2731-8_6

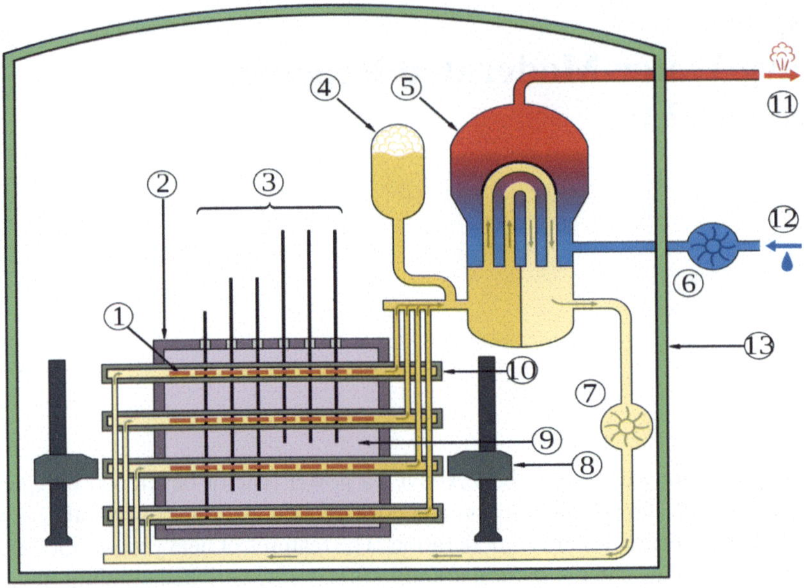

Fig. 6.1 CANDU: Canadian Deuterium Uranium Reactor 1. Fuel bundle, 2. Calandria, 3. Control rods, 4. Pressurizer, 5. Steam generator, 6. Light water pump, 7. Heavy water pump, 8. Fueling machines, 9. Heavy water moderator, 10. Pressure Tube, 11. Steam going to turbine, 12. Cold water returning from turbine, 13. Containment building made of reinforced concrete Reprinted from https://en.wikimedia.org/wiki/CANDU_reactor#/media/File:CANDU_Reactor_schematic.svg under the creative commons attribution-share Akile 2.5 Generic license

which thermally insulate the heavy water in the calandria tank and the pressure tube. The flow of primary cooling water inside the pressure tube is connected at both ends in a loop shape so that it is opposite to the adjacent pressure tube. The fuel rods are uranium oxide fuel pellets, and they use a zirconium alloy (Zircaloy 4) cladding tube with a graphite coating on the inner surface to withstand sudden output changes and stress corrosion cracking during fuel replacement during operation. The materials for the calandria tube and the pressure tube are zirconium 2 and zirconium niobium alloy.

When using natural uranium as fuel, it is necessary to frequently replace the fuel in order to maintain the criticality of the reactor over a long period of time. Therefore, in CANDU, fuel replacement is carried out during operation by a fuel replacement machine. Two fuel replacement machines placed at both ends of the calandria tank insert new fuel from one side of a pressure tube and receive used fuel pushed out from the other side. Out of the 12 fuel assemblies in the pressure tube, 4 each, a total of 8 are replaced at once. CANDU does not need to stop for fuel replacement. Initially, natural uranium was used as fuel, but later, slightly enriched uranium is being used. Since the slightly enriched uranium fuel assembly can be used in the reactor for a longer period of time than the natural uranium fuel assembly, the proportion of fuel processing costs in the power generation cost relatively decreases, and the fuel cycle cost is reduced by 20%–30%. Control rods are inserted from the top into the calandria tank. The configuration of the primary cooling system is similar to that of a pressurized light water reactor, except that heavy water is used as a coolant, and a pressurizer, steam generator, and primary cooling water pump are provided. However, the materials of the piping and equipment of the primary cooling system and others are stainless steel in the pressurized light water reactor, but in CANDU, water treatment is performed on the primary cooling water, and cheaper carbon steel is used. There are two loops in the secondary cooling system. CANDU has the advantage of being able to replace fuel during operation, but it is necessary to compensate for the degradation and loss of heavy water, and the cost of producing heavy water offsets the advantages (IAEA 2023e, ATOMICA 2011).

CANDU reactors, with electrical outputs ranging from 500–800MW, are in operation in Canada. Since the output can be increased or decreased simply by increasing the number of pressure tubes in a CANDU reactor, CANDU3 with 300MW and CANDU9 with 900MW were developed, but they were not built. CANDU uses heavy water as the primary cooling water, and a heavy water production facility is required to compensate for its degradation and leakage. Due to the cost of producing heavy water and in order to reduce construction costs, the ACR1000, which cooled the primary system with light water, was developed from the late 1990s, but its development was discontinued when the nuclear energy division of Atomic Energy of Canada Limited (AECL) was privatized in 2011. When using light water for cooling, the loss of heavy water is reduced, and the volume of the reactor core required is about half due to the moderating effect of light water. However, in order to make the reactivity coefficient negative in the event of a loss of coolant, it is necessary to include a neutron absorbing material in the fuel rods in the center of the fuel assembly (Chan 2002). In light water reactors, when the density of the coolant decreases in the event of a loss of coolant accident, moderation is also lost and a negative reactivity is introduced, but in light water-cooled reactors with heavy water or graphite moderation, moderation is not lost. The neutron absorption rate of the hydrogen in light water is greater than that of the deuterium in heavy water, so if no measures are taken, the reactivity in the event of a loss of coolant will be slightly positive.

The issue of positive reactivity coefficients in heavy water-moderated, light water-cooled reactors was also a problem in the MAPLE-X10 reactor built by AECL for the production of medical radioactive isotopes. This reactor was built as a successor to the NRX reactor of ACEL, which has been supplying medical isotopes to the Western world for many years, but it was found in pre-use inspections that the power reactivity coefficient was slightly positive during startup. Due to the high cost of modifications, AECL gave up on operating the MAPLE-X10 in 2005 (Norquay 2009). Canada produces and supplies radioactive isotopes to the world. In CANDU reactors, cobalt-60, a radioisotope, is produced using adjusters to control the neutron flux distribution in the core, covering most of the world's demand.

Even if the power reactivity coefficient is slightly positive, the power increase rate is small in a state where the nuclear fission chain reaction is maintained, including delayed neutrons. However, when asked if it can respond to all conceivable events, it must answer whether it has exhausted all events, which is impossible. Therefore, since the Chernobyl accident, it has been established as a global safety requirement that the power reactivity coefficient of a reactor must always be negative. This requirement must be met in all operating states from startup to rated operation. At startup, the density of light water in the coolant is high, so it tends to have a more positive reactivity. The size of the reactivity coefficient is influenced by the size of the reactor core, which is related to the leakage of neutrons from the core. The larger the core volume, the smaller the leakage, so the power reactivity coefficient tends to be positive.

This problem is an issue in light water-cooled reactors where the moderator is heavy water or graphite. In light water reactors, the hydrogen in light water slows down neutrons, so the increase in output brings about a decrease in neutron slowing due to a decrease in the density of light water, so it can be designed to always have a negative reactivity coefficient when the coolant boils and the density of light water decreases. The reactivity coefficient of a reactor is related to the size of the neutron cross-section of the moderator and coolant species, the slowing and absorption of neutrons, and leakage from the reactor (the size of the reactor core). Understanding the reactivity coefficient requires advanced knowledge of nuclear physics and reactor design. In exercises in nuclear physics and nuclear engineering, it is the most difficult C level. Even in light water-cooled reactors with heavy water or graphite moderation, design measures are possible, for example, as already mentioned, by placing neutron absorbing material (in this case, Dy: dysprosium) in the fuel assembly like the ACR, the effect of the decrease in neutron absorption when the density of the coolant light water decreases can be relatively small (Chan 2002, p.10). However, the enrichment of the fuel will be slightly higher.

Canada's CANDU reactors had eight units that were shut down for a long period from 1995 to 1998. The reasons were that the records of minor modifications up to that point had not been passed on, and the handover of operational experience from the power company to the reactor design company was poor, causing the maintenance standards to fail and the maintenance costs to rise. This was a typical problem of knowledge succession. A hearing was held to investigate the problem and consider improvements. Six of the eight units were refurbished and restarted by 2003, but the loss due to the long-term shutdown was significant. Two units were decommissioned. In the major refurbishment, the calandria tubes, steam generators, and instrumentation and control systems were replaced (WNA 2024b).

Light water reactors require a nuclear reactor pressure vessel. However, companies capable of manufacturing thick and large steel vessels like nuclear reactor pressure vessels are extremely limited worldwide. Pressurized Heavy Water Reactors (PHWRs) have the feature that they can be manufactured even in countries without such companies. India has introduced CANDU from Canada and is using domestically produced PHWRs. India's first CANDU unit, introduced from Canada in 1972, had a design electrical output of 200MW but operated at 90MW due to technical problems. However, after producing plutonium in the heavy water-moderated research reactor CIRIUS, built with Canada's cooperation, and conducting a nuclear test in 1974, India could no longer receive nuclear cooperation from countries around the world, including Canada. The second unit was a copy of the first unit and was being designed at the time. India then completed the second unit on its own and put it into operation in 1981. Based on this experience, India has developed PHWRs with outputs of 202, 490, and 630MW on its own, and is using a total of 19 PHWRs for power generation. In addition to this, India has two BWRs (electrical output 150MW, both started operation in 1969) and two VVER1000s (started operation in 2013 and 2016) in operation. Civil nuclear cooperation between India and Canada was resumed in 2010.

In addition to these, there are heavy water-moderated boiling light water-cooled pressure tube reactors and pressure vessel type heavy water-moderated and cooled reactors. The former includes the UK's steam generating heavy water reactor (SGHWR, electrical output 100MW, operational period 1967–1990) and Japan's "Fugen" (electrical output 165MW, operational period 1979–2003). SGHWR and "Fugen" are prototype reactors, but no commercial power plants have been built following them. The cooling system of the SGHWR and Fugen is a direct cycle that generates steam to drive the turbine in the core, like a BWR. "Fugen" was operated with a large load of uranium–plutonium mixed oxide fuel (MOX fuel). However, it has not been commercialized because a full MOX ABWR, which can operate with a full core of MOX fuel, has been developed. The full MOX ABWR was being built by J-Power in Oma Town, Aomori Prefecture, but construction was halted when it was about 50% complete due to the TEPCO Fukushima accident in 2011, and work is being done to comply with the new regulatory standards.

The latter pressure vessel type heavy water-moderated reactor was developed by Germany and exported to Argentina, including ATUCHA 1 (electric output 362MW, started operation in 1974) and ATUCHA 2 (electric output 745W, construction started in 1981, commercial operation started in 2014). The reactor pressure vessel houses the core and heavy water, but the flow paths of the heavy water moderator and the heavy water cooling the fuel are separated. The fuel assemblies are each housed in separate cooling channels, with heavy water coolant flowing through them and the heavy water moderator slowly circulating around the outside. The warmed heavy water moderator is used for secondary system feedwater heating and is cooled. The heavy water coolant that has flowed through the fuel channels is sent to the steam generator (IAEA 2022b). In Argentina, a CANDU reactor (electric output 600MW) exported by Canada has been in operation since 1984. The construction of ATUCHA 2 was long interrupted due to Argentina's economic collapse, but construction was resumed with the help of Canada and it started operation in 2014.

References

ATOMICA (2011) Canadian Heavy Water Reactor (CANDU Reactor) ATOMICA, January 2011 (in Japanese). https://atomica.jaea.go.jp/data/detail/dat_detail_02-01-01-05.html. Accessed 12 Aug 2024

Chan P (2002) Evolution of ACR Physics from CANDU 6, AECL Technologies Inc. Presented to US Nuclear Regulatory Commission Office of Nuclear Reactor Regulation December 4, 5, 2002 at CRL. https://www.nrc.gov/docs/ML0300/ML030020092.pdf. Accessed 12 Aug 2024

IAEA (2022b) National Nuclear Safety Report, Argentina, ANNEX III, Main Technical Features of the Argentine Nuclear Power Plants in Operation, 2022, IAEA. https://www.iaea.org/sites/default/files/24/02/national_report_argentina_ar1.pdf. Accessed 12 Aug 2024

IAEA (2023e) Country Nuclear Power Profiles, Canada, IAEA,2023. https://cnpp.iaea.org/public/countries/CA/profile/preview. Accessed 12 Aug 2024

Norquay G (2009) A rock and a hard place: the isotope conundrum, Policy options, IRPP, September1, 2009. https://policyoptions.irpp.org/magazines/employment-insurance/a-rock-and-a-hard-place-the-isotope-conundrum/. Accessed 12 Aug 2024

WNA (2024b) Nuclear Power in Canada, Reactor refurbishment, World Nuclear Association, February 2024. https://world-nuclear.org/information-library/country-profiles/countries-a-f/canada-nuclear-power.aspx. Accessed 12 Aug 2024

Chapter 7
Graphite-Moderated Reactors

The graphite-moderated power reactors include the graphite-moderated carbon dioxide-cooled reactor, the graphite-moderated helium-cooled reactor (high-temperature gas reactor), and the graphite-moderated boiling light water-cooled pressure tube type reactor (RBMK). Graphite has a low neutron absorption rate and can cause a nuclear fission chain reaction even with natural uranium. It is easy to handle because it is solid, and it is easy to manufacture because it is used in industries other than nuclear power. The reactor that Enrico Fermi caused a sustained nuclear fission chain reaction for the first time in the world was also a graphite-moderated reactor. Graphite-moderated reactors were then used as reactors to produce plutonium for nuclear weapons in the United States, the United Kingdom, France, and the former Soviet Union. Based on this experience, the UK and France made carbon dioxide-cooled power reactors. France then introduced light water reactors and built many, while the UK has been using the improved graphite-moderated carbon dioxide-cooled reactor for power generation. The issue of positive coolant void reactivity coefficient of the graphite-moderated boiling light water-cooled pressure tube type reactor such as RBMK is also discussed.

7.1 British Graphite-Moderated Carbon Dioxide-Cooled Reactors

The British graphite-moderated carbon dioxide-cooled reactors are the MAGNOX reactors, which started operation from 1956 to 1971, and the Advanced Gas-cooled Reactor (AGR), which started operation from 1976 to 1988. These reactors were built in pairs (two pairs of four units at two sites) at a single power station site. All MAGNOX reactors (26 units at 11 power stations) have already been shut down. As of 2024, 8 of the 15 built AGRs are in operation, but they are scheduled to be shut down sequentially. In the UK, a PWR has been in operation since 1995. All

Y. Oka, *Nuclear Power Reactor Development*,
https://doi.org/10.1007/978-981-96-2731-8_7

these operating reactors are owned and operated by Electricité de France (EDF)'s UK company (EDF Energy). In addition to this, in UK there was a SGHWR that stopped in 1990 and sodium-cooled fast experimental reactors and fast prototype reactors that stopped in 1977 and 1994. Since 2018, a large pressurized water type light water reactor EPR is under construction. There are also plans to build light water reactors at multiple sites.

MAGNOX covers natural uranium metal fuel with a magnesium alloy, cools the reactor core with carbon dioxide, creates steam with a steam generator (heat exchanger), and sends it to a power turbine to generate electricity. The nuclear reactor pressure vessel was made of steel, but the last four units of the two power plants are made of prestressed concrete. AGR uses enriched uranium oxide fuel, uses stainless steel for the cladding tube, and the nuclear reactor pressure vessel is made of prestressed concrete. The temperature of the carbon dioxide coolant is 600C, and the pressure is 30bar (29.6 atmospheres).

In the UK, the first nuclear research institute (AERE, Atomic Energy Research Establishment) was established in Harwell in 1946, and a graphite-moderated research reactor was in operation since 1948. In 1950 and 1951, two graphite-moderated military plutonium production reactors were in operation at Windscale, facing the Irish Sea. This reactor is air-cooled. In 1953, the government policy to start civilian nuclear power generation was issued, and in 1954, the UK Atomic Energy Authority (UKAEA, United Kingdom Atomic Energy Authority) was established to develop nuclear reactors for power generation. Since a high-temperature coolant is necessary for power generation, carbon dioxide was used for cooling instead of air. The nationalization of the Suez Canal by Egypt in 1956, the tightening of oil supplies, and the soaring prices pushed the strengthening of the nuclear power generation plan. The industry was invigorated, and many heavy electrical manufacturers, boiler makers, and construction companies formed groups and started a race to win orders for nuclear power plants.

The reason why the UK developed graphite-moderated reactors is because they can use natural uranium, eliminating the need for uranium enrichment facilities, graphite is easier and cheaper to produce than heavy water, and they believed they could utilize their experience with graphite-moderated reactors that were operating for research and military plutonium production. In 1956, the first commercial nuclear power plant, the MAGNOX reactor, began operation near Windscale at Calder Hall. This reactor had two purposes, military plutonium production and power generation, but it was later optimized for power generation and increased output. The electrical output was 50MW at Calder Hall, the first MAGNOX reactor, but it increased to about 10 times that at the last Wilfa, reaching 490MW. The power generation efficiency has also improved from the initial 22–28%. The design life was 20 years, but many operated for more than twice that, over 40 years. The MAGNOX fuel cladding tube is made of magnesium alloy, making it difficult to store spent fuel in the air for a long period of time, which is a challenge in the UK's decommissioning measures. The graphite moderator is also left sealed inside the reactor vessel in decommissioning measures for radioactive decay. The operation of many MAGNOX power plants was undertaken by the Central Electricity Generating Board (CEGB), while the UKAEA

Fig. 7.1 BWR and Magnox (Graphite moderated Co$_2$ cooled reactor) Japan atomic Power Company, Tokaimura, Japan *Source* JAPC

was responsible for nuclear research and development, including the operation of the initial eight MAGNOX reactors and the development of fast reactors. MAGNOX was exported to Japan and Italy, but it could not compete with light water reactors due to its high power generation cost. At the Tokai Power Station of Japan Atomic Power Company, a large boiling water reactor (electric output 1100MW) is being built adjacent to this MAGNOX reactor (electric output 166MW). This is shown in Fig. 7.1. The MAGNOX is on the right and the BWR is on the left. The electrical output is about seven times different, but the size of the reactor building (the tall square building in the photo) is almost the same.

In MAGNOX, neutrons are slowed down by collisions with the carbon atomic nuclei of graphite. In light water reactors, they are slowed down by collisions with the hydrogen atomic nuclei of water. The atomic mass of carbon is 12, and that of hydrogen is 1. The proportion of energy lost in a single collision is inversely proportional to the square of the atomic mass. Therefore, graphite is 144 times less efficient than light water in slowing down neutrons, requiring a larger reactor core volume. This is why the reactor buildings of MAGNOX and boiling water light water reactors, which have a seven-fold difference in output, have roughly similar volumes. Construction costs increase with the size of the reactor. Furthermore, gas, being a gas, has a heat removal performance an order of magnitude worse than water, which is a liquid, resulting in larger heat transfer areas (dimensions) for fuel rods and steam generators (heat exchangers). Although the coolant temperature of a gas-cooled reactor is higher than that of a light water reactor, the thermal efficiency (power generation efficiency) is only improved by about 20–30 percent relatively. These are the reasons why the power generation cost of a graphite-moderated gas-cooled reactor is higher than that of a light water reactor. In fact, the power generation cost of the Tokai No. 1 unit (MAGNOX) was higher than that of the Tokai No. 2 unit (BWR). Heavy water-moderated and cooled reactors (CANDU) also have a larger reactor core volume than light water reactors because heavy water is the moderator, and the power density is a fraction of that of light water reactors.

In the UK, the expansion of nuclear power generation was planned in 1964. While CEGB advocated for American light water reactors, UKAEA pushed for Advanced Gas-cooled Reactors (AGR), which were eventually adopted. AGR was operated by UKAEA with an electrical output of 30MW from 1963. As was the case with MAGNOX, the design and construction of power plants in the UK was carried out by consortia of private companies. In the case of AGR, three different consortia of private companies won the contracts for the design and construction of AGRs in various parts of the UK. Both MAGNOX and AGR were developed by the government. Since CEGB, which placed the orders, was a public corporation, it had to consider fairness in the orders, which resulted in a proliferation of manufacturers, with five companies at one time. This led to the dispersion of human resources and investment, and the choice of the AGR route of gas-cooled reactors instead of light water reactors, along with the fact that the export market was dominated by American light water reactors, is said to have been the reason why competitive nuclear reactor manufacturers did not grow in the UK.

The first AGR (electric output 545MW) was designed to achieve a better evaluation compared to light water reactors by incorporating new designs such as fuel exchange during operation and a nearly 20-fold increase in output from the prototype reactor. Due to the lack of experience of the consortium that proposed and won the contract, there were continuous problems during construction, resulting in delays. Construction started in 1965, but commercial operation did not start until 1985, taking 20 years. Fuel exchange during operation has been withdrawn. The first consortium to win the contract went bankrupt, and another consortium took over the construction. In 1973, the National Nuclear Corporation (NNC) was created by the UK government and CEGB, and the remaining two consortia involved in the construction of AGR were integrated.

The issue of AGR construction delays sparked a debate in the UK about the choice of reactor type. MAGNOX, AGR, the High-Temperature Gas Reactor (HTR), Steam Generating Heavy Water Reactor (SGHWR), CANDU, and PWR were considered. The prototype SGHWR (electric output 100MW) has been operating at UKAEA's Winfrith research institute since 1967. CEGB pushed for the American PWR, but in 1975 the British government decided to build two SGHWR power plants. However, the cost of the SGHWR project increased and the commercialization plan was suspended. In 1979, when the government changed to the Conservative Party, the government became positive about the construction of American-made PWRs, and after two years of public hearings, the construction of a large PWR from Westinghouse was decided in 1983. However, due to the development of the North Sea oil field, economic liberalization, power liberalization, the impact of the Chernobyl accident, etc. only one unit (Sizewell B power station, electric output 1188MW) was operated, and for a long time, no nuclear power plants were built in the UK. In recent years, due to the depletion of the North Sea oil field and responses to climate change issues, light water reactor nuclear power plants are now being built.

CEGB's nuclear power business became Nuclear Electric due to liberalization, and then through British Energy, it is now operated by EDF Energy, the UK subsidiary of Electricité de France (EDF). UKAEA's manufacturing division was split into British Nuclear Fuels Ltd. (BNFL), a nuclear energy company such as the nuclear fuel cycle, and a radioactive isotope company in 1971. UKAEA's research division was privatized in 1996 and became AEA Technology, but it withdrew from the nuclear business in 2004. Nuclear fusion research at the Culham Research Institute continues as UKAEA. In 2005, the Nuclear Decommissioning Authority (NDA) was established. The NDA has taken over the business related to UKAEA and BNFL and is conducting nuclear fuel cycle business including decommissioning measures. BNFL has divided and transferred its subsidiary businesses and ended its role in 2010 (WNA 2016, Saido 1993 Part 3 UK).

When comparing the history of nuclear reactor development and commercialization in the UK with that in the US, lessons can be learned from the development by the country and the subsequent commercialization. UK power companies wanted to use US light water reactors, but in the UK, the country was developing prototype reactors, and the country's (government's) intentions were reflected in the choice of reactor type. In the US, private companies have been actively involved in the construction and operation of test reactors for power generation. Even in the construction of the Clinch River Breeder Reactor, a prototype fast reactor, private power companies participate, and the construction plan is evaluated and a decision to stop construction is made depending on whether or not they will make additional investments in response to increased construction costs. The checking mechanisms of the US Government Accountability Office and the Federal Congress are also functioning. US national laboratories may provide their sites for the construction and testing of test reactors, but they do not design or construct prototype power reactors themselves. Since researchers at national laboratories are not designers and the laboratories do not have factories to manufacture equipment, this is rational. The era when the UK developed MAGNOX and AGR was a time when socialism was still strong in the UK. Whether or not power companies spend money is the strictest evaluation of construction plans.

However, under the environment of economic liberalization and power liberalization, there are issues on how to deal with the problem of investment recovery of new construction and the problem of developing new power reactors. Currently, the UK is trying to deal with the problem of new construction investment under the environment of economic liberalization. To use MAGNOX and AGR for many years, it must have been necessary to deal with various unique issues such as material problems. The UK's efforts to operate and use its own gas-cooled power reactors for many years are commendable.

7.2 High-Temperature Gas-Cooled Reactors

The High-Temperature Gas Reactor (HTGR) is a graphite-moderated helium-cooled reactor that aims to achieve a gas temperature of about 750 °C, which is impossible with carbon dioxide cooling, by using helium, an inert gas, as a coolant. To withstand use at high temperatures, small grains of ceramic fuel material are coated with layers of carbon and silicon carbide, embedded in a graphite powder matrix, and compressed and sintered into a spherical or rod shape to be used as fuel. The former is called pebble bed type fuel, and the latter is called prism type fuel; the former is used in Germany and China's HTGR, and the latter is used in the US, the UK, and Japan's HTGR. The coated fuel particles, unlike the metal cladding tubes of light water reactors, cannot completely contain volatile fission products (FP) at high temperatures. This is not a safety issue, but in HTGR, care must be taken to adhere FP to pipes and generate radioactive dust during maintenance of primary cooling system equipment.

High-temperature gas test reactors were built in the United States, the United Kingdom, and West Germany in the 1960s. These include the UK's Dragon (1966–1975, thermal output 20MW, outlet temperature 750 °C, pressure 20bar, steel nuclear reactor vessel), the US's Peach Bottom Unit 1 (1967–1974, electrical output 40MW, coolant outlet temperature 750 °C, pressure 25bar, steel nuclear reactor vessel), and West Germany's AVR (1968–1988, electrical output 15MW, outlet temperature 950C, pressure 11bar, steel nuclear reactor vessel). In addition, small test reactors were also built in the 1960s at places like the Oak Ridge National Laboratory and the Idaho National Reactor Testing Station in the United States. Like other power reactors, HTGRs generate steam using steam generators and generate electricity with steam turbines, but because the outlet temperature is high, the power generation efficiency is about 40%, which is relatively 20% higher than that of light water reactors. Early small test reactors also experimented with using gas turbines in a closed cycle for power generation.

Following the high-temperature gas test reactors, the United States and West Germany built prototype reactors for commercial HTGRs. These include the Fort St. Vrain reactor, which began construction in the United States in 1968 and went into operation in 1979 (US, electrical output 330MW, coolant inlet temperature 405C, outlet temperature 775 °C, pressure 48bar, prestressed concrete reactor vessel (PCRV)), and the THTR300 (West Germany, electrical output 300MW, outlet temperature 750 °C, pressure 20bar, PCRV) (Kupitz 1984, Loftness 1964). The Fort St. Vrain reactor was decommissioned in 1989 due to problems with the helium circulator, among other things, and was converted into a gas turbine combined cycle power plant in 1996. The THTR also had problems such as damage to the pebble bed type fuel and the associated problems, and damage to the bolts of the high-temperature duct, and only operated for a short period from 1985 to 1989.

In Japan, the high-temperature helium gas cooled test research reactor HTTR (thermal output 30MW, outlet temperature 850–950 °C, pressure 4MPa) became critical in 1998 and achieved an outlet temperature of 950 °C in 2004. The HTTR is not for power generation, but for research on high-temperature heat utilization such as hydrogen production.

China has been operating the test reactor HTR10 (electrical output 10MW, outlet temperature 700–950C), which was built based on West Germany's AVR, since 2003. After that, based on this experience, they built the Shidao Bay Power Plant (electrical output 200MW, helium outlet temperature 750C, secondary system steam maximum temperature 560C, pressure 7MPa) as a commercial power reactor. Construction began in 2012, and although it was slightly delayed, Unit 1 is in operation in 2021. China does not consider HTGR as an alternative to light water reactors, and applications such as replacement of coal-fired power plants in suburban areas and heat utilization have been proposed (Zhang 2017). Since the commercial HTGR prototype reactors in Europe and the United States are not operating smoothly, the operation status of China's HTGR is attracting attention.

The disadvantage of graphite-moderated reactors having high construction costs due to low power density is offset by the safety feature of the reactor core having a large heat capacity and slow behavior during loss of cooling accidents. Even if cooling is lost, the reactor output naturally decreases. There are nuclear experts who evaluate this point. This feature is related to how the national regulatory agency judges the safety of the reactor, including the need for evacuation during an accident, which investor to invest in, and how to consider the relationship between technical safety and the public's concerns. As mentioned in the Atomic Energy Act, safety regulations are necessary because technology has benefits to use, not just for nuclear power. If there are no benefits, it should be prohibited. Therefore, the issues of safety and regulation are fundamentally judged by comparing costs and benefits. The benefit would be to provide cheap and stable electricity to the public.

A technology is used in the market only after it wins competition with competing methods. In thermal utilization, the superiority or inferiority with competing non-nuclear technologies becomes an issue. For example, the gas temperature obtained by burning fossil fuels exceeds 1000 °C, which is not achieved by the maximum temperature of helium gas in HTGR. In hydrogen production, the method of making hydrogen from water using nuclear heat is in competition with methods such as decomposing ammonia. As a method for seawater desalination, the competitiveness of seawater desalination by nuclear heat utilization is raised as an issue against the widely used reverse osmosis method.

There is a regulatory issue for commercial high-temperature gas reactors as to whether a nuclear reactor containment vessel is necessary. The commercial power plants that have operated in the United States are light water reactors and HTGR's Peach Bottom Unit 1 and Fort St. Vrain. Only Fort St. Vrain of these adopts a confinement instead of a containment that is airtight. The containment is one of the three barriers to prevent the release of nuclear fission products into the environment, along with the fuel cladding and primary coolant boundary, and is an element of the basic principle of nuclear safety, the concept of defense in depth. Confinement refers to a low-pressure containment vessel that is pressure-releasing, and is kept slightly negative than atmospheric pressure during normal operation. In the event of a depressurization accident such as a rupture of the primary cooling pipe, the helium coolant released into the vessel is released into the atmosphere by opening the valve, and then the valve is closed to confine the nuclear fission products. The concept of confinement takes advantage of the fact that the behavior of HTGR during an accident is slow, and it takes time for the coated particle fuel to become hot and release gaseous fission products.

Fort St. Vrain was built in the 1970s, and it is necessary to consider knowledge and experience gained since then. The Electric Power Research Institute (EPRI) in the United States published a report in 2005 that examined the design, regulation, location, and economic aspects of containment, and raised issues. "Design issues include measures against aircraft collisions and tests of pressure relief valves. In terms of regulation, the containment vessel has a unique function to control the release of radioactive substances into the environment that are not present in the fuel or cooling system when an unexpected event occurs. For example, it is effective in reducing the impact of air intrusion accidents in HTGRs, so it is necessary to consider whether the confinement can perform the same function, and to examine the pressure resistance of the vessel and the re-closure ability of the vent path. It is premature to consider conditions that can reduce the area where emergency plans are required. More knowledge of air ingress accidents is needed. Data such as tests on the retention of fission products in coated particle fuel after irradiation is also needed. Consideration of aircraft collisions is also necessary. It is necessary to consider events that have a large environmental impact in events outside the design basis (referred to as cliff-edge events)." (Loflin 2005). The High-Temperature Engineering Test Reactor (HTTR) in Japan has a containment vessel.

7.3 Graphite-Moderated Boiling Light Water-Cooled Pressure Tube Reactors: RBMK

The Soviet Union's Graphite-Moderated Boiling Light Water-Cooled Pressure Tube Reactor (RBMK) is a power generation version of the graphite-moderated light water-cooled reactor used for the production of plutonium for nuclear weapons. The plutonium production reactor that the United States operated in Hanford, Washington, during World War II was a graphite-moderated light water-cooled reactor, cooled by atmospheric pressure water, and used natural uranium. The Soviet plutonium production reactor was of the same type. The plutonium nuclear bomb of the United States was dropped on Nagasaki in 1945, but the Soviet Union succeeded in nuclear explosion in 1949. The Soviet nuclear power generation plan began the following year in 1950. To use this plutonium production reactor for power generation, the Soviet Union pressurized light water, used enriched uranium as fuel, and built a reactor called APS-1 in Obninsk, about 100 km southwest of Moscow. This was the test reactor for the RBMK, a graphite-moderated boiling light water-cooled pressure tube power reactor, which was the first in the world to succeed in nuclear power generation in 1954. The thermal output was 30MW. The electrical output is

5MW. The inlet and outlet temperatures of the cooling water are 190C and 280C, and the pressure is 10MPa. The cladding tube is a double concentric cylinder made of stainless steel, and the fuel elements are filled with grains of uranium–molybdenum metal with a concentration of 5% along with a magnesium matrix between the outer and inner tubes. Four fuel elements and one cooling tube are provided in a graphite matrix, constituting a single fuel channel (length 6.775m). The core is composed of 128 fuel channels and 23 control rod channels. APS-1 was the first to generate nuclear power in the world, but it has since been used more for district heating than for power generation (Kotchetkov 2004).

Based on APS-1, two prototype reactors of RBMK nuclear power plants (with electrical outputs of 108MW and 160MW) were built in Beloyarsk and operated for about 20 years from the late 1960s. Beloyarsk had a nuclear superheater (a part in the core that produces superheated steam, superheated steam is high-temperature steam that has further heated the saturated steam generated by boiling water) in both Units 1 and 2, but subsequent RBMK1000s do not have a superheater and use saturated steam for power generation.

The RBMK is composed of equipment that can be manufactured in existing factories, and does not require thick and heavy structures like the pressure vessel of a light water reactor or complex equipment like a steam generator, and therefore does not require special manufacturing facilities. It has features such as being able to increase the output of a nuclear power plant just by increasing the number of channels, being able to replace fuel during operation, and so on. The RBMK1000 with an electrical output of 1000MW started operation in 1974, and a total of 16 units have been built. Most power plants are built in pairs of two, with a common turbine building. In addition, two RBMK1500s with an electrical output of 1500MW have been built. The RBMK1500 had the world's largest electrical output at the time. It is said that the construction cost was reduced by 20–30% compared to the RBMK1000 (Semenov 1983). The RBMK has been operated for about 50 years after undergoing major repairs after the Chernobyl accident and to cope with aging, but it is gradually being replaced by the light water reactor VVER.

The core and cooling system of the RBMK are shown in Fig. 7.2. As shown, the cylindrical core is composed of numerous fuel channels made of graphite moderator, fuel assemblies, and cooling tubes. Control rods are inserted from the top of the core. Numerous cooling tubes (outer diameter 88mm) exiting the core are connected to a steam separator, where the core cooling water is separated into steam and water. The steam is sent to a steam turbine for power generation. The separated water is mixed with the condensate returning from the steam turbine and is pumped into numerous cooling tubes at the bottom of the core by a recirculation pump. There is no steam generator, and the cooling system is a direct cycle configuration that generates steam for power generation directly in the reactor, similar to a boiling water light water reactor or SGHWR.

The core is housed in a cylindrical steel reactor vessel (inner diameter 9.7m, height 14.52m, thickness 16mm). The radial walls of the vessel are double-layered, and nitrogen gas is filled between them. The upper and lower ends of the inside and outside of this vessel are bellow structures, which can absorb the difference in thermal

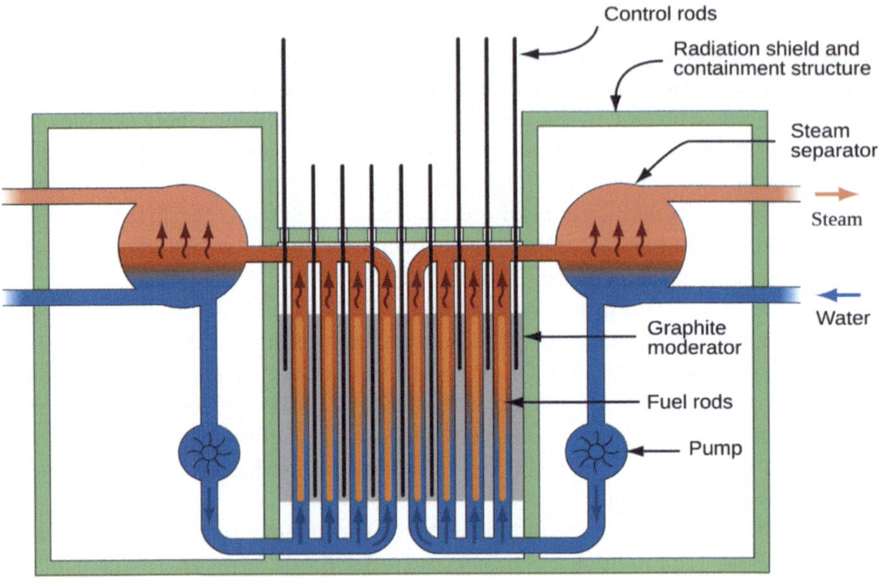

Fig. 7.2 Schematic diagram of RBMK Reprinted from https://en.wikimedia.org/wiki/RBMK#/media/File:RBMK_reactor_schematic.svg under CC BY-SA 3.0

expansion between the inner and outer vessel walls. The graphite moderator blocks are stacked inside the reactor vessel to a diameter of 8m and a height of 14m, and are placed in an inert atmosphere of a mixture of helium and nitrogen. On the radial outside of the reactor vessel, there is a double-cylindrical water tank, and a layer of sand is located between it and the outer reinforced concrete wall. This reinforced concrete wall, along with the layers of water and sand, serves as a radiation shield. In the reactor cooling water, radioactive nitrogen-16 is produced by the reaction of high-speed neutrons and oxygen nuclei in the core. Although it has a short half-life of 7 s, it emits strong gamma rays. Therefore, the steam separator and recirculation pump are also covered in a radiation shielding building. There is radiation shielding at the top and bottom of the reactor vessel. Since the upper shielding is penetrated by pipes for the fuel channels and measurement channels, an upper shielding cover is provided above it, which serves as the floor of the reactor building. Fuel replacement is performed by removing the plug from this floor.

The cooling system consists of two independent systems, each responsible for half of the fuel channels in the reactor core. Each cooling system is equipped with two steam-water separators and four recirculation pipe pumps (one of which is on standby). The cooling water sent from the recirculation pump is fed into the nuclear fuel channels through 22 distributors. The inlet temperature of the cooling water in the reactor core is 265–270 °C, the outlet temperature is 284 °C, and the pressure is 6.9MPa, which is almost the same as a boiling water light water reactor. The steam sent to the power turbine becomes water at 155–165 °C in the condenser and enters the liquid phase of the steam drum. The water separated from steam in the steam drum is saturated water, so it is 284C. In RBMK, when the reactor output is low, the balance of the flow rate of the separated water and the water returning from the condenser tends to be disturbed, and the inlet temperature of the reactor core tends to be unstable. Signals for scram (reactor shutdown) are set to be issued in case of high and low water levels in the steam drum, high steam temperature, low feedwater flow rate, and stoppage of the recirculation pump, but these signals could be canceled by the operator. In boiling water reactors, it is required that the ratio of output to flow rate is within a certain range when starting up.

The emergency core cooling system, when the reactor is under high pressure, has two pressurized systems and a system that injects water from the vent using a pump. The latter can be driven by residual steam from the turbine when power for the pump cannot be obtained. Long-term cooling has three water injection systems by electric pumps, using water from the pressure suppression pool of the containment vessel as a source. Power for the pump can also be obtained from emergency diesel generators.

The fuel rods of the RBMK1000 are made by sealing hollow fuel pellets of uranium oxide (diameter 11.5mm, height 15mm, hollow part diameter 2mm) in a zirconium alloy cladding tube (outer diameter 13.6mm, thickness 0.825mm), with an enrichment of about 2%. The fuel assembly piece bundles 18 fuel rods around a central support rod (diameter 13mm) in a circular shape. The fuel assembly piece is 3.64m long, and two pieces are combined vertically to form one fuel assembly. The total length is 10.025m, of which the effective fuel part is 6.862m. There is a fuel-free area in the center of the fuel assembly. Fuel replacement can be done during operation. The fuel replacement machine is set above the channel where the fuel assembly to be replaced is located, the pressure is made the same as inside the reactor, the spent fuel assembly is removed, and new fuel is loaded. The effective core height of the RBMK is about 7m and the diameter is 11.8m. This is about 12 times larger in core volume than a boiling water light water reactor of almost the same output, with a core height of about 3.76m and a diameter of 4.66m. The reason for this is that, as mentioned in the section on gas-cooled reactors, the slowing down ability of graphite is smaller than that of light water.

The control rods of the RBMK are inserted from the top of the reactor core. The lower end of the control element is connected to a part made of boron carbide, a strong neutron absorber, through a region filled with 1.25m of water and a 4.5m graphite follower. The absorption of neutrons is strongest in the boron carbide part, followed by the water part, and weakest in the graphite part. The effective height of the RBMK core is 6.862m, so when all the control rods are withdrawn, the graphite part comes to the center of the core, with a 1.25m water region on both sides. From this state (with the control rods withdrawn), when the control rods are inserted, the water region of the control rod channel at the bottom of the core is first replaced with graphite, reducing neutron absorption and introducing a positive reactivity into the reactor. This is a phenomenon known as the "positive scram" of the RBMK, which is said to have been a cause of the Chernobyl accident. In the RBMK reactor, the water region also served as a damper, so the driving speed of the control rods was slow, about 0.4m/s. In the Chernobyl accident, due to the effect of the positive scram and the slow driving speed of the control rods, the output suppression effect was delayed, the reactor output increased for a long time, and the output increased significantly before the effect of the control rods being inserted was demonstrated. Furthermore, due to the positive reactivity output coefficient unique to the RBMK, the output of the core increased rapidly, the nuclear fuel became high temperature and granulated, reacted rapidly with water, causing steam explosion and destroying the reactor. After the accident, the RBMK has been modified to increase the insertion speed of the control rods.

The positive scram was a trigger, and the positive reactivity coefficient was the fundamental cause of the Chernobyl accident. The problem of the positive reactivity coefficient can be improved by design. For example, by adding a neutron absorber to the fuel, which will slightly increase the fuel enrichment, if its effect is superior to the reduction of neutron absorption due to boiling of light water, the reactivity coefficient can be made negative.

The positive reactivity power coefficient unique to RBMK is a characteristic of nuclear reactors that use graphite or heavy water as moderators and light water as coolant. This was mentioned in the section on heavy water reactors, but here we will explain it in terms of graphite-moderated reactors. When the light water coolant boils and its density decreases, the neutron absorption by hydrogen, which constitutes light water, decreases, increasing the reactivity of the nuclear reactor and raising its output. In light water reactors, the effect of reducing neutron moderation by light water, which decreases reactivity, outweighs the effect of reducing neutron absorption by hydrogen, and reactivity decreases. However, in graphite-moderated reactors, the moderation effect hardly changes, and the reactivity increases because the neutron absorption effect by hydrogen decreases. In other words, the temperature coefficient of reactivity (power coefficient) becomes slightly positive. As a result, when light water boils, the output of the nuclear reactor increases. At low output, the coolant is cold and dense. Therefore, the temperature coefficient of reactivity tends to become more positive. When starting a nuclear reactor and increasing its output, if there is an output range with a positive reactivity power coefficient, the output will increase on its own, making control not easy. It seems that the RBMK, which has such characteristics, was not easy for operators to start or operate at low output. Not only for RBMK, but for all nuclear reactors, the power reactivity coefficient needs to be designed to be negative throughout the entire power range from zero to rated power. As shown in the example of Canada's ACR, it is possible to design it ingeniously and make the reactivity power coefficient negative throughout the entire power range.

At the time of the Chernobyl reactor accident, a test was being conducted to see if residual steam for power turbines could be used as the driving force for the emergency core cooling system pump. The nuclear reactor was at low output, some safety systems were made inoperable, and the test was conducted without permission. However, due to the instability of the RBMK at low output, the operators struggled to realize the experimental conditions and could not control the nuclear reactor. The test procedure was also not well examined or checked. The report of the IAEA's International Nuclear Safety Advisory Group (INSAG) points out that the design weaknesses of the RBMK were revealed in other RBMK accidents in 1975 and 1982, and that the positive scram was revealed in the RBMK1500 in 1983, and emphasizes the importance of safety culture (IAEA 1992).

RBMK does not have a containment vessel like a light water reactor that houses the reactor and primary cooling system. This is because the volume of the reactor core is large due to graphite moderation, and a huge airtight building (containment vessel) is required. The length of the RBMK fuel assembly is about 2.5 times that of a light water reactor, and at least five times the height of the containment vessel of a light water reactor is required to pull it out. In the case of RBMK, it was thought that safety could be ensured by following the country's rules even without a containment vessel, while the construction cost and construction period of a nuclear power plant would double if a large containment vessel was made. Because there was no containment vessel, the upper part of the reactor and the cooling pipes there were not covered by an airtight structure. In the Chernobyl accident, the upper structure of the reactor was blown away by a steam explosion in the reactor core, and the shattered high-temperature core fuel and materials were exposed to the air. The high-temperature graphite came into contact with the air and caught fire, and radioactive substances in the fuel were scattered into the air with smoke (NEA 2002, pp.11). After the Chernobyl accident, partial containment facilities were built for some RBMK reactors (Wikipedia 2024). The explosion of the reactor building in the TEPCO Fukushima accident was different from the explosion of the Chernobyl reactor, as it was caused by hydrogen generated by the reduction of water by high-temperature fuel that had lost cooling, which remained in the reactor building and explosively burned with oxygen in the building. The environmental contamination of the TEPCO Fukushima accident was mainly due to volatile fission products (such as cesium-137), while in the Chernobyl accident, fuel materials in the reactor core were also scattered by the fire in addition to volatile radioactive substances.

References

IAEA (1992) INSAG-7, The Chernobyl Accident: Updating of INSAG-1, IAEA Safety Series No.75, 1992. https://www-pub.iaea.org/MTCD/Publications/PDF/Pub913e_web.pdf. Accessed 12 Aug 2024

Kotchetkov L (2004) Obninsk, Number One, Nuclear Engineering International, July 13, 2004. https://www.neimagazine.com/advanced-reactorsfusion/obninsk-number-one/. Accessed 12 Aug 2024

Kupitz J, Dee JB (1984) International status of HTGRs, IAEA Bulletin, vol.26. No.4, 1984. https://www.iaea.org/sites/default/files/publications/magazines/bulletin/bull26-4/26404780510.pdf. Accessed 12 Aug 2024

Loflin L (2005) Containment Versus Confinement for High Temperature Gas Reactors; Regulatory, Design Basis, Siting, and Cost/Economic Considerations. 1011948 Technical Updates May 2005 EPRI

Loftness RL (1964) Nuclear Power Plants, Chapter 9 High-Temperature Gas Cooled Reactors, Van Nostrand, 1964

NEA (2002) Chernobyl: Assessment of Radiological and Health Impacts, 2002 Update of Chernobyl: Ten Years On, OECD/ NEA, 2002. https://www.oecd-nea.org/upload/docs/application/pdf/2022-01/3508-chernobyl_2022-01-05_11-11-9_843.pdf. Accessed 12 Aug 2024

Saido K, Gray JE (1993) The Miracle of Nuclear Power, Nikkan Kogyo Shimbun, 1993 (in Japanese)

Semenov BA (1983) Nuclear Power in The Soviet Union, IAEA Bulletin vol.25, No.2, 1983. https://www.iaea.org/sites/default/files/25204744759.pdf

Wikipedia (2024) RBMK, Wikipedia. https://en.wikipedia.org/wiki/RBMK. Accessed 12 Aug 2024

WNA (2016) Nuclear Development in the United Kingdom, Nuclear Power in the United Kingdom Appendix 1, World Nuclear Association, October 2016. https://www.world-nuclear.org/information-library/country-profiles/countries-t-z/appendices/nuclear-development-in-the-united-kingdom.aspx. Accessed 12 Aug 2024

Zhang Z (2017) The Status of HTR-PM, a 200MWe High Temperature Gas-cooled Reactor demonstration plant constructed in China, International Ministerial Conference on Nuclear Power in the 21st Century Abu Dhabi, 30 October to 1 November 2017. https://www.iaea.org/sites/default/files/17/11/cn-247-zhang.pdf. Accessed 12 Aug 2024

Chapter 8
Liquid Metal-Cooled Fast Breeder Reactors and Fast Reactor Fuel Cycle

The liquid metal-cooled fast breeder reactor, which can produce more fissile material than it consumes while generating power, was developed as a "dream reactor" in the United States, the Soviet Union, the United Kingdom, France, and later Germany and Japan, along with the beginning of the peaceful use of nuclear power. It is currently used in Russia and is under development in China and India. The first nuclear power generation in the United States was not by a light water reactor, but by a test reactor of a fast reactor called EBR-1 (electric output 200 kW), which was in 1951. There is a photo of four light bulbs glowing.

Natural uranium contains only about 0.7% of fissile Uranium-235. 99.3% is non-fissile Uranium-238. When Uranium-238 captures a neutron, it undergoes radioactive decay to produce fissile Plutonium-239. By reprocessing spent fuel to extract plutonium, and processing it into fuel along with uranium, and loading it into a fast reactor, it is possible to produce more nuclear fuel material (plutonium) than is consumed while generating power.

In nuclear fission, when fast (high-speed) neutrons are collided with fissile plutonium nuclei (Pu239, Pu241) without being slowed down, about three fast neutrons are generated. Of these, one neutron, excluding those that leak from the reactor or are absorbed by atomic nuclei other than nuclear fuel materials, maintains the chain reaction by causing the next nuclear fission reaction, and when one or more of the remaining neutrons are absorbed by Uranium-238 atomic nuclei, they undergo radioactive decay and generate fissile plutonium nuclei. By chemically processing (reprocessing) spent fuel, extracting this plutonium, and mixing it with uranium to make nuclear fuel, it is possible to produce more fuel (fissile plutonium) than is consumed while generating electricity. This is the principle of the fast breeder reactor. If nuclear fission is not performed with fast neutrons, the number of neutrons generated per fission is small, so breeding is not possible when using uranium–plutonium fuel. In fast reactors, light water and heavy water, which have a large neutron slowing effect, are not used as coolant, but liquid metal is used.

Y. Oka, *Nuclear Power Reactor Development*,
https://doi.org/10.1007/978-981-96-2731-8_8

Light water reactors slow down neutrons generated by nuclear fission to make them react more easily with uranium nuclei and cause a chain reaction of nuclear fission. Although plutonium is generated in spent fuel of light water reactors, it is not possible to produce more fissile material than is consumed while generating electricity (while maintaining a chain reaction of nuclear fission). Because slowing down neutrons makes nuclear fission reactions more likely, light water reactors can use lowly enriched uranium as fuel, but fast reactors need to use fuel with a high enrichment (concentration of fissile plutonium) in the nuclear fuel. The plutonium enrichment of fast reactor fuel is 20–30%. Plutonium has a shorter half-life than uranium and does not exist naturally. Reprocessing spent fuel to extract plutonium is essential for achieving breeding in a fast breeder reactor. In a fast breeder reactor, in order to continue generating electricity with the bred fuel, it is required that the fast breeder reactor and the reprocessing facility for spent fuel from the fast reactor operate as a set.

Sodium, with a mass number of 23 and low neutron slowing ability, is mostly used as a coolant for fast breeder reactors. Sodium has a high melting point of 98C, but NaK is liquid at room temperature. It was used in the UK's fast experimental reactor DFR. However, NaK is chemically, highly reactive. Special treatment is required in the decommissioning measures of DFR. In the UK, liquid sodium was used as a coolant in the fast prototype reactor PFR, which was built later. The former Soviet Union had fast reactors using lead–bismuth alloy (melting point 125C) as a coolant for submarines until the 1970s, but the reactors of submarines have been replaced with pressurized water light water reactors. Lead–bismuth alloy does not react as violently as liquid sodium when it comes into contact with water or air. Lead-cooled fast reactors have also been proposed, but lead–bismuth alloy has a lower melting point than lead (melting point 327.5C). When using sodium, lead–bismuth alloy, or lead as a coolant, it is necessary to keep the coolant pipes warm even when the reactor is stopped to prevent freezing. Liquid metal-cooled reactors are low pressure and have a liquid surface inside the reactor vessel, but the upper part of the liquid surface needs to be covered with an inert gas. If air enters the cover gas, the liquid metal is oxidized, and the oxide adheres to the cooling system, so it is necessary to prevent this.

Figure 8.1 shows a system diagram of a loop-type liquid metal-cooled fast reactor (LMFBR). Fast reactors in the United States and Japan are loop type, but those in the UK, France, and Russia are a pool type, which has an intermediate heat exchanger installed inside the reactor vessel. In either type, the heat of the liquid sodium (primary sodium) that cooled the reactor is exchanged with the secondary sodium through the intermediate heat exchanger, and steam is generated using a steam generator, which is then sent to a steam turbine to generate electricity. The pool type does not require insulation of the primary sodium piping, and measures against leakage of sodium from the primary piping are not necessary, so it is systematically simple. The loop type is said to be easy to maintain the primary pump and piping, but a floor cover is needed to prevent the chemical reaction between the floor concrete and the radioactive liquid sodium when it leaks. The construction cost is slightly lower for the pool type.

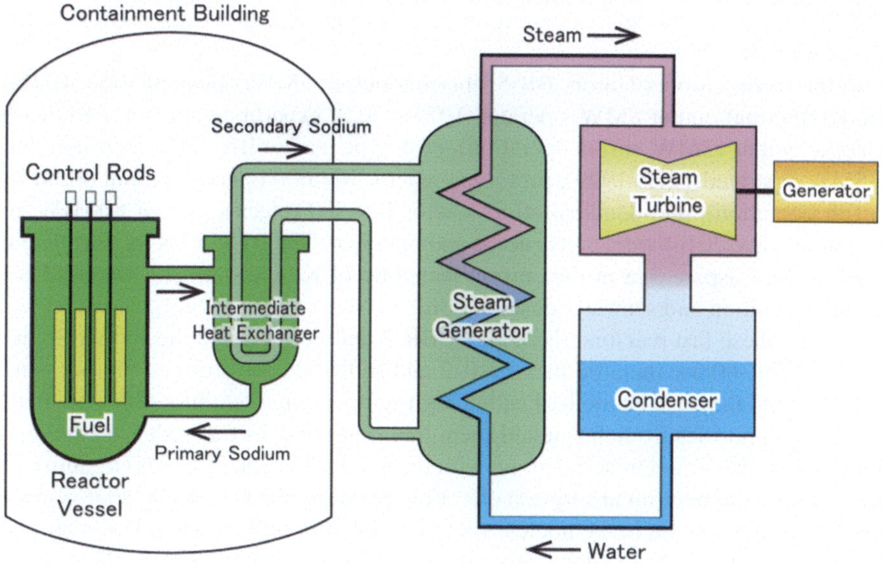

Fig. 8.1 Liquid metal cooled fast breeder reactor (loop-type) Reprinted from https://commons.wikimedia.org/wiki/File:Fast_Breeder_Reactor_01.svg under the creative commons attribution 4.0 international license

The primary sodium that cooled the reactor is radioactive, so if it is directly led to the steam generator and heat-exchanged with water, there is a risk of radioactive sodium reacting with water and scattering if a leak occurs in any of the numerous heat transfer tubes of the steam generator. To prevent this, both the pool type and the loop type use an intermediate heat exchanger to transfer the heat of the primary sodium to the secondary sodium, and heat exchange with water in the steam generator to make steam. The secondary sodium is not radioactive, so even if it reacts with water, there is no risk of radioactive substances scattering. The steam generator is equipped with a valve that opens the pressure to the atmosphere when a sodium-water reaction occurs due to a leak in its heat transfer tube, and a cyclone separator to collect the reaction products. This equipment can be seen, for example, in exterior photos of the "Monju" building.

In the United States, following the operation of EBR-1 (electric output 200kW, operated from 1951 to 1963), EBR-2 (electric output 20MW, operated 1963–1994), Fermi-1 (electric output 61MW, operated 1963–1975), SEFOR (thermal output 20MW, operated 1969–1972), and FFTF (Fast Flux Test Facility, thermal output 400MW, operated 1980–1994) were constructed and operated. The U.S. fast breeder prototype reactor Clinch River Breeder Reactor (CRBR) had its budget canceled in 1983 due to soaring construction costs and construction delays. In the U.S., President Carter opposed reprocessing and fast breeder reactor plans in 1977 due to concerns that reprocessing technology could potentially lead to nuclear proliferation, partly due to India's nuclear test in 1974. The International Nuclear Fuel Cycle Evaluation

(INFCE), an international discussion on the entire nuclear fuel cycle, was held for two years from 1977.

In the former Soviet Union, BR-5 (thermal output 5MW, operated 1958–1971), BR-10 (thermal output 8MW, operated 1973–2002), experimental reactor BOR-60 (electric output 12MW, still in operation), prototype reactor BN-350 (electric output 135MW, operated 1972–1999), prototype reactor BN-600 (electric output 600MW, still in operation), and demonstration reactor BN-800 (electric output 800MW, in operation since 2014) were constructed and operated. The BN-350 is located on the coast of the Caspian Sea in the current Republic of Kazakhstan and was used for power generation and seawater desalination.

Among these fast reactors, the U.S.'s EBR-2 and Fermi-1 and the former Soviet Union's BOR-60 use metallic fuel. EBR-2 and BOR-60 have attached reprocessing facilities, and they used a method called high-temperature metallurgy (dry method) to remove impurities from the melted spent fuel, and filled the fuel rods with the melt containing nuclear fuel material along with the residual fission products (fissium). In fast reactors, the neutron absorption effect of fission products is weak, so it is not a problem if they remain in the nuclear fuel material after reprocessing. However, the radiation dose is still high after reprocessing, so it is necessary to handle the process from reprocessing to fuel processing remotely. The main fuel for other fast reactors is oxide fuel. They are treated by dissolving fuel pellets in nitric acid, which is used in light water reactor reprocessing (wet method). In the case of the wet method, the fast reactor fuel has a high plutonium concentration, so it needs to be mixed with spent light water reactor fuel to lower the plutonium concentration before dissolving it in nitric acid. This reprocessing technology is an extension of light water reactor fuel reprocessing technology. In the wet method, since fission products are removed, the processing of plutonium obtained by reprocessing into fuel can be done in a glove box under negative pressure relative to atmospheric pressure, not remote operation with radiation shielding. The reason for handling in a glove box is to prevent workers from inhaling plutonium. There is no need to process uranium fuel in a glove box.

The American company GE, based on experiences such as EBR-2, proposed a small, integrated, metal-fueled fast reactor called PRISM in the 1980s. The integrated type refers to a design that incorporates an intermediate heat exchanger into the reactor vessel. The pool type has the same configuration, but the term pool type is used for high-output liquid metal-cooled fast reactors such as French Phenix and Superphenix. In PRISM, a power plant consisting of several small-output reactor modules manufactured in a factory was proposed (Salerno 1988).

The modular reactor is intended to reduce construction costs by producing a large number of reactor modules of the same design. A gas turbine combined cycle power plant operates as a single power plant by arranging several gas turbine modules. After PRISM, modular power plants have been proposed for nuclear power generation. Currently, small modular reactors (SMR) are also proposed for light water reactors. There is an opinion that the cost reduction effect of module manufacturing does not show a cost reduction effect in a few units, and it does not show a cost reduction effect unless it manufactures more than several tens of units like a gas turbine.

Gas turbine combined cycle power generation is an innovation in power generation technology, so research has been conducted to apply gas turbine technology to nuclear power generation. Carbon dioxide, which becomes liquid when pressurized and cooled (31.1C or less), was considered as a heat transport medium. Jet engines and gas turbines for thermal power generation are, however, combustion turbines (a type that obtains power by burning and expanding liquid fuel), and are open cycle gas turbines that release exhaust gas into the atmosphere after combustion. In contrast, when using a gas turbine for nuclear power generation, it is necessary to make it a closed cycle to return the gas discharged from the gas turbine to a liquid (in a light water reactor, water is used as a heat transport medium, and the steam is returned to water in a condenser. If it does not return to water, the turbine will not rotate and power generation will not be possible.). A large heat exchanger is required for cooling to return the discharged carbon dioxide to a liquid. Since the gas turbine combined cycle power plant is an open cycle, this heat exchanger is not necessary (although a waste heat recovery boiler for recovering exhaust gas heat is installed as a common facility for multiple turbine generators, it is for improving power generation efficiency and its role is different). The use of gas turbine technology for nuclear power generation has been studied as a concept, but has not been put into practical use.

British fast reactors include the Dounreay Fast Reactor (DFR, electric output 15MW, operated 1959–1977), and the Prototype Fast Reactor (PFR, electric output 250MW, operated 1974–1994). Dounreay is located at the northernmost tip of Scotland, where a fast reactor fuel reprocessing plant and a material test reactor were also placed, and the development of a pressurized light water reactor for British nuclear submarines was also carried out (Jensen 1995).

In the UK, the UKAEA published a review of the UK's fast reactor program in 1987. The report, produced by the UKAEA, which was responsible for the research and development of fast reactors, provides insight into the state of the UK's fast reactor development at the time. It includes information such as the UK government's statement that electricity would bear about 30% of the cost of fast reactor research and development, cooperation with France and Germany's fast reactor programs, the number of employees, and budget related to fast reactors. The status of PFR operation experience, reprocessing, commercial reactor design, structural integrity, fast reactor equipment, materials, chemistry, core and fuel, safety, analysis methods, and calculation programs are described. The report states that there were 920 experts in the field of fast reactors in the UK at the time, 800 of whom were affiliated with the UKAEA (Bramman 1987). No fast reactors have been built in the UK since the PFR. The UKAEA's research division was privatized in 1996.

French Fast Reactor: In France, the Atomic Energy Commission systematically carried out the development of fast reactors and reprocessing technology on a national scale. French fast reactors include the experimental reactor Rapsodie (operated from 1967 to 1983), the prototype reactor Phenix (electric output 250MW, operated from 1973 to 2009), and the commercial reactor Superphenix (electric output 1240MW, operated from 1986 to 1998). These fast reactors were built and operated by the French Atomic Energy Commission (CEA). Phenix and Superphenix are pool-type fast reactors with uranium–plutonium mixed oxide fuel (MOX fuel). The reprocessing of spent fuel from these fast reactors was carried out at the AT1 fast reactor fuel facility in La Hague from 1969 to 1977, and at the APM reprocessing facility in Marcoule from 1974 to 1995. All of these fast reactors and reprocessing facilities are either decommissioned or in the process of decommissioning. Phenix was built in Marcoule and operated at a high capacity factor in the 1970s and 80s, achieving a breeding ratio of 1.16 and demonstrating the breeding performance of fast reactors. The 1990s saw a low-capacity factor due to time spent investigating negative reactivity anomalies and repairing and modifying equipment. Various experiences have been gained from the operation of Phenix as a prototype of a pool-type fast breeder reactor. As for the leakage of the coolant sodium, it has been experienced many times, once a year during power operation, like other countries' fast reactors (Guidez 2013; IAEA 2007).

The commercial fast reactor Super Phoenix was created by making necessary improvements to the large-scale reactor based on the experience of the prototype reactor Phoenix. As a commercial reactor, Electricité de France (EDF) was responsible for its construction and operation. Italian nuclear reactor manufacturers also participated in the design and manufacture. The Super Phoenix had a low operating rate of about 20% due to the time required for the reactivation process due to the handling of malfunction events, the anti-nuclear movement caused by the Chernobyl accident in 1986, and the sodium fire caused by the massive leakage of sodium into the air at a Spanish solar thermal power plant that used liquid sodium as a heat medium in the same year. The malfunction events include sodium leakage from the fuel storage container, degradation of primary sodium due to air intrusion into the sodium purification system, and argon leakage from the container surrounding the intermediate heat exchanger. More time was required for public hearings and reactivation procedures than for repairs and countermeasures.

The IAEA's 2007 report on "Experience in the Design and Operation of Liquid Metal Cooled Fast Reactors" compares the costs of the French 1400MWe pressurized water light water reactor called P'4 and the Super Phoenix, stating that the construction cost of the Super Phoenix is 2.66 times that of P'4 and the power generation cost is 2.22 times (IAEA 2007, page 84, Table 7). The 2012 report titled "Cost of Nuclear Power Generation" by the French Court of Auditors (Cour des Comptes) states that the cost of constructing and operating the Super Phoenix from 1974 to 1997 was 12 billion euros (1.44 trillion yen in 2010 value). In Japanese materials comparing the construction costs of sodium-cooled fast reactors and light water reactors (1981 costs, not the first unit but the fifth unit), it is 2.49 times for loop type and 2.22 times for pool type. Looking at the breakdown, the cost of facilities related to sodium is about five times that of corresponding facilities in light water reactors (Oka 2018, page 104). The construction cost of a sodium-cooled fast breeder reactor is higher than that of a light water reactor due to facilities for using liquid sodium as a coolant, such as cooling system facilities, sodium purification systems, and cover gas systems. The cost evaluation results by the research and development institutions responsible for the research and development of fast reactors are lower than these. Looking at the operating experience, not only sodium leakage, but also issues and points of attention associated with using liquid metal as a coolant, such as the formation and adhesion of sodium oxides due to the intrusion of air into the cover gas system of liquid sodium, have become clear. This is in contrast to light water reactors, which are water-cooled and have many commonalities with the design and operating experience of thermal power generation equipment.

In France, after the Super Phoenix, the role of fast reactors is related to the disposal of high-level radioactive waste, and the separation and conversion of long-lived transuranic elements (Separation and Transmutation, sometimes inaccurately referred to as "incineration of radioactive waste" or "hazard reduction" in Japan) was proposed by a researcher at the French Atomic Energy Commission. In Japan, the Atomic Energy Society of Japan has decided to call it "separation and transmutation". In France, in 1991, the Radioactive Waste Management Research Act (Bataille Act) was established to address the issue of high-level radioactive waste disposal. Three methods—geological disposal, separation and transmutation, and long-term storage—were examined over a period of 15 years, with geological disposal being chosen and efforts being made toward high-level radioactive waste disposal. The French Atomic Energy Commission had hoped to construct a fast prototype reactor called ASTRID, but in a report published by the French National Evaluation Committee in July 2020, there was little mention of fast reactors.

West Germany also developed a fast reactor; the test reactor KNK-2 (electric output 20MW, operated 1972–1991) was designed by Interatom and built and operated at the Karlsruhe Nuclear Research Center. The firstly built KNK-1 was a thermal neutron reactor with zirconium hydride moderation and liquid sodium cooling, but it was shut down in 1974, and seven fuel assemblies in the core were replaced with uranium–plutonium mixed oxide fuel assemblies without zirconium hydride moderation rods, so that fuel assemblies for the fast prototype reactor SNR300, which was planned in West Germany at the time, could be tested. The reactor core was changed and the name was changed to KNK-2. After the modification, KNK-2 started operation in 1978. The reprocessing of KNK-2 fuel was also carried out at the adjacent reprocessing facility using the PUREX method (wet method). The obtained nuclear fuel material containing plutonium was processed at the fuel processing plant in HANAU and loaded into KNK-2. Although it was on a kilogram scale, West Germany also succeeded in closing the nuclear fuel cycle (reprocessing fuel irradiated in a fast reactor, processing it into fuel, and loading it into a fast reactor).

Construction of West Germany's fast prototype reactor SNR300 began in Kalkar on the lower Rhine in 1973, but it was caught in political turmoil. From 1979 to 1982, a study was conducted on whether to continue construction, and it was decided to continue. In 1985, liquid sodium was filled into the primary cooling system, and it was ready to operate in 1986, but the political party of the state where the construction site was located adopted a coal-first policy, and the plan stalled. In 1989, the company responsible for reprocessing suddenly announced its withdrawal from the reprocessing business. In 1991, the Minister of the Federal Ministry of Research, along with the top executives of West Germany's three major power companies and the nuclear reactor manufacturing company Siemens, decided to cancel the SNR300 plan because it seemed unlikely that the state where the construction site was located would grant permission to operate the SNR300. The SNR300 was built but not operated. The experience with fast reactors was supposed to be handed over to the European Fast Reactor (EFR) project, but the budget for West Germany's fast reactors was not continued (Marth 1994). The site of the SNR300 was sold along with the facility in 1995 and is now used as a theme park.

Sodium-cooled fast reactors in Japan include the experimental reactor "Joyo" (thermal output 140MW, operational 1978–) and the prototype reactor "Monju" (electric output 280MW, operational 1994–2010). Both are loop type, and the Power Reactor and Nuclear Fuel Development Corporation (later the Nuclear Fuel Cycle Development Institute, now the Japan Atomic Energy Agency) was responsible for their construction, operation, and management. Joyo is located in Oarai Town, Ibaraki Prefecture, and has been used for irradiation tests of fuel materials for sodium-cooled fast reactors. The thermal output was initially 50MW during performance testing, but it was increased to 100MW and 140MW by adding more fuel to the core. Joyo was shut down in 2007 due to damage to the upper part of the irradiation test experimental device. This device was restored in 2014, but it has been shut down for a long time due to compliance with new regulatory standards established after the TEPCO Fukushima accident (as of April 2023). To set the evacuation plan established after the TEPCO Fukushima accident to a range of 5 km, modifications were made to limit the thermal output to 100MW.

"Monju" began generating power in August 1995. In December, while gradually increasing the output, it experienced a sodium leak and was shut down. It resumed operation in May 2010, but was immediately shut down when a radioactive gas detector malfunctioned. In August, a device for fuel exchange, the in-reactor relay device, fell, resulting in a long-term shutdown. In 2013, the Nuclear Regulation Authority ordered an indefinite shutdown, citing missed inspections of important equipment and false reporting during safety inspections. In 2015, the Nuclear Regulation Authority announced inspection results showing that about half of the improvement items had not been confirmed, and recommended to the Minister of Education, Culture, Sports, Science and Technology to specify an operating entity to replace the Japan Atomic Energy Agency. In December 2016, the government decided to decommission "Monju". The total operational days were 250.

The cause of the sodium leak in December 1995 was due to the "sheath" that housed the thermocouple thermometer measuring the sodium temperature vibrating and breaking due to the flow of sodium. Sodium leaks have occurred frequently in fast reactors overseas, and operations have been resumed after recovery, but in Japan, it has caused a big fuss and led to a long-term shutdown. Failures that do not become a topic of conversation in other industries become a big fuss and attract public attention in the case of nuclear power, especially fast reactors and reprocessing. Experimental reactors and prototype reactors are made to confirm the design and gain experience, so it is difficult to develop new nuclear technologies in a country where every failure becomes a big fuss. Looking at these failures as challenges in the nuclear industry, the author can mention the difficulty in inheriting the design and manufacturing experience of fast reactor equipment in Japanese nuclear reactor manufacturing companies, due to the long period of time it took to construct commercial fast reactors after the construction of "Monju", such as the damage to the "Joyo" irradiation test experimental device and the fall of the "Monju" in-reactor relay device.

Ensuring the safety of the public associated with the use of nuclear power, not limited to fast reactors and nuclear fuel cycle businesses, is the role of the nuclear regulatory agency. Nuclear operators who operate and manage nuclear facilities should always communicate with regulatory agencies. When visiting overseas nuclear facilities, the author has heard from the person in charge of the facility that he talks to the resident regulatory officer for at least 30 min every day. It is also good to follow the rules set by the nuclear regulatory agency regarding the disclosure of information about accidents and failures. If there is a provision in an agreement with a local government, it is necessary to follow that as well. When holding a press conference in the event of an accident or failure, it is desirable for the person in charge to receive training in communication skills in advance. The skill to convey the main points clearly is an essential ability for the person in charge of a nuclear business, not only in response to accidents and failures. Although the cases are different, the way late Yukiya Amano, the former Director General of the IAEA, spoke at general meetings and press conferences was easy to understand and impressive. He mentions in his book that he received training from a British actor. In the case of the sodium leak at "Monju" and the damage to the irradiation test experimental device at Joyo, there are people who committed suicide among the practical charge of the Power Reactor and Nuclear Fuel Development Corporation and the Japan Atomic Energy Agency. There is a philosopher's opinion that "blaming a fire thief on a fire is less appropriate than attributing the cause to theft" and "misleading reports and statements should be ethically questioned" (Ichinose 2022, page 26). Can we attribute the suicide of the person in charge to the sodium leak? Isn't there a common factor with the reputational damage at the time of the TEPCO Fukushima accident?

The development of fast reactor fuel reprocessing technology is positioned as an extension of the development of light water reactor spent fuel reprocessing technology. In Japan, it was planned to be carried out by the Power Reactor and Nuclear Fuel Development Corporation (PNC, now the Japan Atomic Energy Agency). The reprocessing of spent light water reactor fuel was carried out at the PNC's Tokai reprocessing plant, processed into mixed oxide fuel, and operated after being loaded into the new conversion reactor "Fugen". The spent fuel of the fast reactor had a higher plutonium concentration than the light water reactor fuel, so it was planned to add a pretreatment process to the reprocessing plant and dilute the concentration for reprocessing. However, since the Rokkasho Village reprocessing plant of Japan Nuclear Fuel Limited was built, the Tokai Village reprocessing plant of PNC, including the facilities prepared for the research and development of fast reactor fuel reprocessing, is transitioning to decommissioning measures.

India has been promoting fast reactor cooperation with France and has been operating a fast experimental reactor called FBTR since 1985. The electrical output is 13MW. The FBTR is a loop type. The prototype reactor PFBR is a pool type with an electrical output of 500MW and has been under construction since 2004. Its design is said to have been based on the French Phoenix. As of 2024, it is still not in operation. The reasons are thought to be technical problems and the need to comply with strengthened safety regulations due to events such as the Indian Ocean tsunami and the Great East Japan Earthquake. The fast reactor fuel reprocessing and processing facilities are adjacent to these fast reactors.

In China, the experimental reactor CEFR (electrical output 20MW) has been in operation since 2011. The first unit of the prototype reactor CFR-600 (electrical output 642MW) has been under construction since 2017, and the second unit since 2020.

References

Bramman JI, Wheeler RC (1987) A Review of the United Kingdom Fast Reactor Programme, March 1987. https://inis.iaea.org/collection/NCLCollectionStore/_Public/45/100/45100184.pdf. Accessed 12 Aug 2024

Guidez J (2013) Phenix, the experience feedback, CEA, edp sciences, 2013

IAEA (2007) Liquid Metal Cooled Reactors: Experience in Design and Operation, IAEA-TECDOC-1569, December 2007, IAEA. https://www-pub.iaea.org/MTCD/Publications/PDF/te_1569_web.pdf. Accessed 12 Aug 2024

Ichinose M (2022) On nuclear power policy, from the perspective of a philosopher, 25th Japan Atomic Energy Commission Meeting Document, No. 1. June 27, 2022, page 26 (in Japanese). https://www.aec.go.jp/kaigi/teirei/2022/siryo25/1_haifu.pdf. Accessed 12 Aug 2024

Jensen SE, Oelgaard PL (1995) Description of the Prototype Fast Reactor at Dounreay, RAK-2, Riso National Laboratory, December 1995. https://inis.iaea.org/collection/NCLCollectionStore/_Public/28/026/28026107.pdf. Accessed 12 Aug 2024

Marth W (1994) The History of the Construction and Operation of the German KNK II Fast Breeder Power Plant, KfK5456, November 1994. https://publikationen.bibliothek.kit.edu/270038327. Accessed 12 Aug 2024

Oka Y (2018) Challenges and Human Resource Development in Japan's Nuclear Utilization, Japan Professional Engineers Association Nuclear and Radiation Subcommittee Special Lecture Meeting, Machinery Promotion Hall June 15, 2018, (Atomic Energy Commission Homepage Activity Introduction June 13, 2018) (in Japanese). http://www.aec.go.jp/jicst/NC/about/kettei/180613.pdf. Accessed 12 Aug 2024

Salerno LN (1988) Salerno et al PRISM concept, modular LMR reactors. Nuclear Engineering and Design 109 (1988):79–86. https://www.sciencedirect.com/science/article/abs/pii/0029549388901446. Accessed 12 Aug 2024

Chapter 9
Uranium Resources, Uranium Enrichment, Reprocessing, and Plutonium Utilization in LWRs

This chapter discusses uranium resources, uranium enrichment, reprocessing, and the use of plutonium in light water reactors. First, we will introduce two perspectives on uranium resources and will answer the question: Will uranium resources be depleted? Uranium enrichment is necessary to supply enriched uranium to light water reactors. Gaseous diffusion methods were commercialized first in the United States and then in France, followed by centrifugation in Europe and Russia. It required less electricity for enrichment. It was the innovation of the technology. Reprocessing is the process of treating spent fuel to extract plutonium. The use of plutonium obtained through reprocessing as fuel for light water reactors was first carried out in the United States and later in Europe.

After World War II, the United States set the goal of commercializing nuclear power and developing fast breeder reactors. At that time, the amount of uranium resources discovered was extremely small. Nuclear fuel materials were managed by the U.S. government, and it was thought that they would be insufficient due to military use. This is the background to the U.S. setting the goal of developing fast breeder reactors at that time. The uranium enrichment facility was made in Oak Ridge, Tennessee, during World War II, and enriched uranium was also managed by the U.S. government. Countries without uranium enrichment facilities developed graphite-moderated or heavy water-moderated power reactors using natural uranium as fuel. However, the U.S. later began supplying low enriched uranium to the world, and many light water reactors were built in countries other than the U.S.

9.1 Uranium Resource

There are two perspectives on the amount of uranium resources in the world. One is the view that uranium resources will eventually deplete, and the other is that uranium, as a metal resource, is distributed at various concentrations on the earth like other

Y. Oka, *Nuclear Power Reactor Development*,
https://doi.org/10.1007/978-981-96-2731-8_9

metal elements, and if mining costs are not spared, it will not deplete. The former is often cited with evidence compiled by mining companies and others, which includes confirmed reserves and estimated reserves, categorized by mining costs. In the past, this was compiled by the Uranium Institute in the UK, but since the 2000s it has been announced by the World Nuclear Association (WNA), which was reorganized and established from the Uranium Institute. The WNA is a private organization with nuclear-related companies as members. Currently, it is compiled and announced by the OECD/NEA.

Figure 9.1 shows the distribution of uranium in the earth's crust (Monnet 2016). If you take the grade of uranium on the horizontal axis and show the amount of uranium, it is distributed in a bell shape (bell curve) like other metal elements. The uranium resources currently being mined are in the high-grade part. The confirmed reserves and estimated reserves compiled based on information from mining companies and others, which are considered as uranium resource amounts, are part of this high-grade part. If the grade drops by one digit, the resource amount increases 300 times. Uranium is also dissolved in seawater. Research on seawater uranium collection has been conducted in Japan, but it has not been put into practical use due to the cost of recovery.

Fossil fuels originate from ancient organic matter. Some people seem to think that uranium resources will deplete by analogy with fossil fuels of organic origin, but the worry of depletion is relative, and if it really becomes necessary, there is plenty in nature, and recovery technology and economics become key. For example, looking

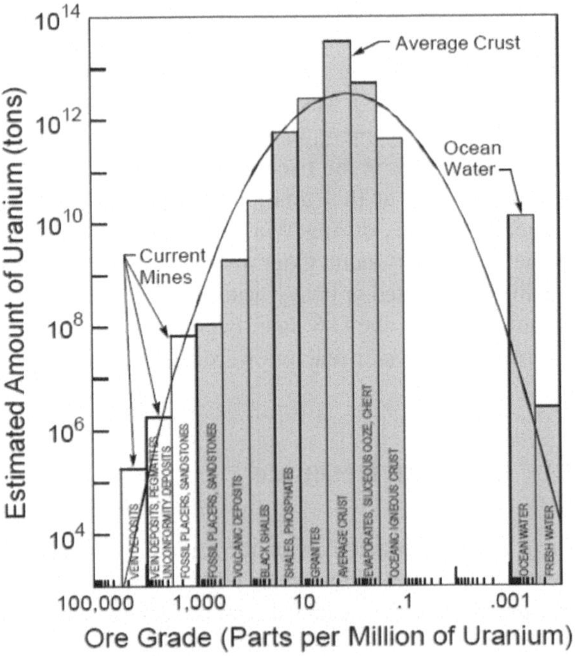

Fig. 9.1 Uranium distribution in mined deposits and earth's crust *Source* A. Monnet, S. Gabrie and J. Percebois "Statistical model of global uranium resources and long–term availability ", EPJ Nuclear Sci. Technol., 2 (2016) 17

at the history of oil exploration and mining technology, it is clear that its recovery technology has advanced. In the 1960s, a technology was developed to drill horizontally under the seabed of the Gulf of Mexico, increasing oil production. Natural gas was also produced in conjunction with oil, but with the use of technology to drill horizontally and fracture shale with hydraulic pressure to recover the contained natural gas and oil, the production of natural gas (shale gas) increased as demand for natural gas increased. With the increase in production, natural gas prices in the United States have fallen, and the cost of natural gas power generation has been decreasing in the 2010s. Not limited to oil and natural gas, fluctuations in resource prices occur due to the balance of supply and demand, and there is no direct relationship with the amount of resources.

Metals are ubiquitous in nature, and their resources are unlimited if cost is not a concern. When demand exceeds supply, prices soar, which drives exploration and inspires the development of recovery technologies, increasing the amount of extractable reserves and causing prices to plummet. This cycle repeats itself. Uranium resources are economically sustainable, just like other metal resources. While oil and natural gas resources are concentrated in specific countries such as the Middle East and Russia, uranium resources are not. They are produced widely around the world, including in Canada, Australia, Kazakhstan, Russia, and Africa, and are not concentrated in politically unstable regions like oil. Historically, uranium prices have risen around the 1970s when many light water reactors were being built worldwide, and around 2010, which was referred to as the nuclear renaissance. In the long term, the price of uranium ore is decreasing.

9.2 Uranium Enrichment

To use uranium as fuel for light water reactors, it is necessary to enrich Uranium-235, which only makes up 0.7% of natural uranium, to between 3 and 5%, creating low enriched uranium. Uranium-238, which makes up 99.3% of natural uranium, and Uranium-235 have almost no difference in their chemical properties, so gas diffusion and centrifuge methods are used for enrichment. Power companies purchase refined uranium and request enrichment services from companies that own enrichment plants.

The gas diffusion method, developed during World War II, enriches uranium by taking advantage of the slight difference in diffusion speed between Uranium-235 and Uranium-238. A gaseous uranium compound (uranium hexafluoride, with a vaporization temperature of 57C) is sent into a gas chamber with a partition with many small holes, and the slightly greater amount of Uranium-235 hexafluoride that passes through is used for enrichment. This is done in multiple stages. Uranium enrichment by gas diffusion consumes a large amount of electricity. Large-scale uranium enrichment by gas diffusion was carried out in the United States and France. In the United States, it was carried out at enrichment plants owned by the U.S. Department of Energy in Oak Ridge, Tennessee, and Paducah, Kentucky. The K-25 plant in Oak

Ridge operated during World War II until 1983. Several enrichment plants have been built in Oak Ridge following K-25. The Paducah enrichment plant operated from 1952 to 2013 and was the only enrichment plant in the United States from 2001 to 2013. Until the 1970s, the United States was the only country outside the communist bloc that had commercial uranium enrichment facilities and supplied enriched uranium. France, in the 1970s, along with Italy, Belgium, and others, established Eurodif and built a uranium enrichment plant using the gas diffusion method in their country, which operated from 1979 to 2011. The gas diffusion method has the advantage of being able to adjust enrichment according to changes in demand, but it consumes a large amount of electricity. Currently, all gas diffusion enrichment plants have been shut down, and the centrifuge method is being used.

In the centrifuge method, Urenco, a company established in the 1970s by the UK, the Netherlands, and Germany, has started commercial uranium enrichment. Urenco also has an enrichment plant in New Mexico, USA. France has built a centrifuge plant based on Urenco's technology as a successor to the gas diffusion plant and has been operating it since 2011. Russia began introducing the centrifuge method in the early 1960s and has been exporting low enriched uranium to Western countries since the 1970s. As of 2020, the largest enrichment capacity is held by Russia's Rosatom, followed by Urenco, France's Orano, and China's CNNC. All of these use the centrifuge method. It is difficult to adjust the operation rate of the plant according to demand in the centrifuge method. Therefore, the world's enrichment capacity is significantly exceeding demand (WNA 2022). Therefore, the price of enriched uranium is stable.

In Japan, Japan Nuclear Fuel Ltd., a company established with investment from power companies, has been operating an enrichment plant in Rokkasho Village, Aomori Prefecture, since 1992. The scale is small. Uranium enrichment is a delicate technology related to nuclear non-proliferation. In Japan, the technology developed by the Power Reactor and Nuclear Fuel Development Corporation is being used by Japan Nuclear Fuel Ltd. through technology transfer. In addition to this, although on a small scale, Argentina, Brazil, India, Pakistan, and Iran are enriching uranium. After the end of the Cold War, the United States purchased 500 tons of highly enriched uranium from dismantled Russian nuclear weapons, down-blended it into low enriched uranium, and used it for nuclear power generation, but this supply has largely ended.

The amount of electricity required for the centrifuge method is significantly less, about 2%, compared to the gas diffusion method. The centrifuge method is a technological innovation in enrichment technology. The cost of uranium enrichment accounts for about half of the cost of nuclear fuel. Other costs include fuel fabrication costs and uranium ore costs. The proportion of enrichment costs in the generation cost of light water reactors is about 5%. The fuel cost of nuclear power generation is not a large proportion of the fuel cost of uranium ore because the proportion of enrichment costs and fabrication costs is large. This is in contrast to the generation cost of thermal power generation, where the proportion of fuel costs for natural gas and coal is large.

Uranium enrichment methods include gas diffusion and centrifugal separation, which have been commercialized, but other methods such as laser, nozzle, and ion exchange have also been researched. The laser method excites the isotope (Uranium-235) with a laser to separate it. There are two types of laser methods: atomic and molecular. The Sirex method, a combination of laser atomic and nozzle methods, was invented in Australian uranium enrichment research and has a high separation efficiency. In the 2000s, GE in the United States tried to commercialize it, but gave up due to the oversupply and downturn in the uranium enrichment market. Currently, SILEX in Australia is aiming for commercialization with investment from a Canadian company. In Japan, the Laser Enrichment Technology Research Association was established in the late 1980s, and atomic laser enrichment technology development was carried out by researchers from nuclear reactor manufacturers for about 20 years. The molecular method was researched by RIKEN and the Japan Nuclear Fuel Limited, and the ion exchange method (chemical method) was researched by a private chemical company in Japan.

9.3 Reprocessing

Reprocessing refers to the extraction of plutonium from spent fuel for reuse. There are broadly three reprocessing methods, wet, pyrometallurgical, and electrolytic, but currently, the method of dissolving spent fuel in nitric acid, known as the PUREX method, a type of wet method is used. Commercial reprocessing facilities around the world include La Hague in France (processing capacity 1700t/y), Mayak in Russia (400t/y), Rokkasho in Japan (800/y), and Sellafield in the UK (1500t/y), which processes spent fuel from MAGNOX reactors and LWRs. There are three plants in India that reprocess spent fuel from heavy water-moderated pressure tube reactors (total 300/y). The Rokkasho Reprocessing Plant, which started construction in 1993, has not yet started operation as of April 2025. The Rokkasho Reprocessing Plant is a massive construction project equivalent to five or six light water reactors. In Japan, the Japan Nuclear Fuel Limited operated a reprocessing plant (capacity 210 t/y) in Tokai Village since 1977, but instead of scaling up this plant, Japan Nuclear Fuel Limited built most of the reprocessing process with the cooperation of a French company based on the design of the UP2-800 reprocessing plant (capacity 800 t/y) that was operating in France. However, the domestically produced glass solidification facility (a facility that dissolves and solidifies radioactive waste liquid generated in reprocessing with glass) connected to it did not operate well, and it took a long time to modify and obtain approval. The Tokai reprocessing plant of JAEA was decommissioned in 2018.

Japan Nuclear Fuel Limited (JNFL) is conducting reprocessing, uranium enrichment, and low-level radioactive waste disposal operations in Rokkasho Village, Aomori Prefecture. In Aomori Prefecture, JNFL is a large-scale company that far surpasses others, contributing to the employment of the prefecture's residents. Despite past issues surrounding the development of Mutsu Ogawara area, JNFL is a private company, and the responsibility for the reprocessing business lies with JNFL and the power companies. This is the same as the responsibility for the TEPCO Fukushima accident falling on TEPCO. In the long term, responding to changes in the focus of nuclear power utilization should be beneficial not only for companies but also for the prefecture and its residents if the development of this region is to be pursued. First and foremost, it is important to expand the interim storage capacity of spent fuel.

In the UK, a reprocessing plant called THORP, which reprocesses spent fuel from AGR and light water reactors (capacity 900t/y), was in operation from 1994 to 2018 at Sellafield. In Japan, the Japan Atomic Energy Agency operated a reprocessing plant (capacity 210t/y) in Tokai Village, Ibaraki Prefecture, from 1977 to 2006, reprocessing spent fuel from light water reactors and the prototype conversion reactor "Fugen". The capacity of the reprocessing plant is the maximum possible capacity, which differs from the actual processing volume. The United States operated a 300t/y light water reactor spent fuel reprocessing plant from 1965 to 1972 in West Valley. A 1500t/y plant was built in Barnwell, but it could not operate due to a change in President Carter's reprocessing policy. Other reprocessing plants currently in operation around the world include a military one in Tomsk, Russia, and smaller ones in China, Israel, and Pakistan.

9.4 Plutonium Utilization in LWRs

In the fuel cycle of light water reactors, there is the once-through cycle, which does not reprocess spent fuel, and the plutonium recycle, which reprocesses spent fuel and uses the resulting plutonium to make uranium–plutonium mixed oxide fuel, which is then loaded into light water reactors for power generation. The latter is called "pluthermal" in Japan because it uses plutonium in a thermal neutron reactor, which is a light water reactor. This is a Japanese-made English term. Even though it is called recycling, in light water reactors, the isotopic composition of plutonium deteriorates after several cycles, affecting the nuclear characteristics of the light water reactor, so only one cycle of recycling is performed. Currently, it is systematically carried out in France's 900MW class light water reactors. The history is old, and it was carried out in Belgium in 1963, and it used to be carried out in light water reactors in Germany and other European countries, but it gradually stopped, and now it is not carried out in Europe except in France. As a result, the UK's oxide fuel reprocessing plant has been closed. The high burnup of uranium fuel and the once-through cycle are the mainstream of light water reactor fuel use in the world and are used in countries other than France.

Japanese power companies have entrusted the reprocessing of spent fuel to the UK and France, and have had the resulting plutonium processed into uranium–plutonium mixed oxide fuel, which is then brought back to Japan and loaded into their respective light water reactors. The first loading took place in 1986. In recent years, several power companies have been delaying their "pluthermal" plans because it is not easy to obtain the consent of local governments, resulting in the accumulation of their reprocessed plutonium in the UK and France. The plutonium generated by German and Swedish power companies, which have also entrusted reprocessing to the UK and France, has already been consumed and is no longer present. Since the Japanese power companies have already completed their reprocessing contracts with the UK and France, there is no risk of an increase in the amount of Japanese plutonium in these countries. However, it is undesirable for Japan, a non-nuclear-weapon state, to accumulate large amounts of plutonium, in terms of promoting nuclear non-proliferation worldwide.

As time passes after reprocessing, americium accumulates in plutonium, degrading its characteristics as a fuel. Therefore, it is better to match the amount of reprocessing to the consumption of plutonium to prevent its accumulation. This was stated as a policy of French reprocessing in a report on the compliance status of the "Convention on the Safety of Spent Fuel Management and on the Safety of Radioactive Waste Management" submitted by the French government to the International Atomic Energy Agency (IAEA) in 2003. This policy seems to have been unknown in Japan. This is an example of biased information being circulated in the Japanese nuclear community, and should be a lesson for the future. Such situations can be avoided by careful searching in English.

The Japan Atomic Energy Commission has decided on the "Basic Concept of Plutonium Use in Japan" in 2018. It states that "reprocessing should be done according to the amount consumed in 'pluthermal', reducing the amount of plutonium held, power companies should cooperate in reducing the amount of plutonium held overseas, and efforts should be made to expand the storage capacity of spent fuel". It should be noted that even in light water reactors with uranium fuel, plutonium is generated in the fuel and it undergoes nuclear fission to generate energy, so there is not much difference between "pluthermal" and light water reactors with uranium fuel. Japan's prototype converter reactor "Fugen" has operated with more than 50% of its core fuel being MOX fuel (uranium–plutonium mixed oxide fuel), and has used 772 MOX fuel assemblies to date.

The U.S. Department of Energy began construction of a MOX fuel processing plant at Savannah River in 2005 to consume plutonium generated from the dismantling of nuclear weapons in U.S. light water reactors. However, construction was halted due to soaring construction costs and the inability to find many power companies willing to use MOX fuel.

References

Monnet A et al (2016) Statistical model of global uranium resources and long-term availability. EPJ Nuclear Sci Technol 2:17. https://doi.org/10.1051/epjn/e2016-50058-x

WNA (2022) Uranium enrichment (updated October 2022), World Nuclear Association. https://world-nuclear.org/information-library/nuclear-fuel-cycle/conversion-enrichment-and-fabrication/uranium-enrichment.aspx. Accessed Aug 12 2024

Chapter 10
Economic and Political Aspects of Plutonium Use

This chapter discusses the economics and political aspects of plutonium utilization.

Light water reactors and uranium enrichment are competitive fields with many international companies involved, but reprocessing has only ever been carried out by British and French companies, and currently only French companies are conducting commercial reprocessing, so there is no competitive environment. Therefore, there is little information available about the cost of reprocessing. The cost of electricity generation by France's Plutonium–Uranium Mixed Oxide (MOX) fuel is said to be only slightly higher than that of uranium fuel. The report "The Cost of the Nuclear Power Sector" issued by the French Court of Auditors (Cours des comptes) in January 2012 states on page 34 the past investment amount of AREVA facilities. According to this, about half of the investment in AREVA's reprocessing and enrichment facilities was covered by overseas customers. This is thought to have reduced the cost burden of France's MOX fuel. On the other hand, a group of nuclear non-proliferation experts at Harvard University in the United States stated in a 2003 report that MOX fuel would not have the same electricity generation cost as the once-through cycle of uranium fuel unless the uranium price reached $360/kgU from the current $40/kgU. They conservatively adopted $1000/kgHM as the reprocessing cost and $1500kgHM as the MOX fuel fabrication cost (Bunn 2003).

Since the price of uranium is only a small part of the electricity generation cost, comparing only the price of uranium does not fully reflect its impact on the electricity generation cost, but it is certain that the cost of MOX fuel is higher. The cost of Japan's MOX fuel could be significantly increased by its reprocessing cost and MOX fuel processing cost. There is sometimes debate about the feasibility of the reprocessing project in Rokkasho Village, but this issue is primarily a matter of management for Japan Nuclear Fuel Limited and the electric power companies. Japan's nuclear fuel reprocessing and MOX fuel are carried out by private companies, the electric power companies. The responsibility lies primarily with the power companies. It should be better in the end not to leave responsibility ambiguous.

Y. Oka, *Nuclear Power Reactor Development*,
https://doi.org/10.1007/978-981-96-2731-8_10

In a report from Harvard University, comparisons are also made with the fast reactor cycle, but in this case, it is believed that the sensitivity of the construction cost of the fast reactor to the power generation cost is high. It may not be easy to overcome more than twice the construction cost difference to light water reactors in research and development for liquid metal-cooled fast reactors. Large-scale research and development investments should be made with a view to solving problems, learning from past experiences. The accountability of administrative agencies (responsibility for the results of exercising budgets and authority) should also include making use of past facts and lessons. If we do not explore the issues, we cannot get the opportunity to solve them, so it is not that we are saying that research on fast reactors is unnecessary. However, it is a mistake to ignore past lessons. Especially in the case of universities, the outputs are students who have acquired knowledge and research papers, and they do not require a large amount of research and development budget, so what to research, including fast reactors, is free. The outputs of research and development institutions are systematized knowledge, software, and experimental results for use, and a large budget, including the employment cost of researchers, is required to produce these, and accountability is questioned.

Fukushima Prefecture has been holding the "Energy Policy Review Meeting" since June 2001, and has discussed policies such as "pluthermal (plutonium utilization in LARs)". The interim summary was announced in September 2002. The meetings were held 35 times until 2005 and an international symposium was held in this year. The review meeting was resumed in 2009, and the 39th held in the last 2010; the verification results related to the interim summary were reported. At the 31st held in 2004 (Bunn 2003), a US nuclear non-proliferation expert gave a lecture, and the minutes (summary version) and materials are also open to the public. The presentation of this nuclear non-proliferation expert is based on the above-mentioned 2003 Harvard University report (Bunn 2003). From these minutes, you can understand not only the thinking of US nuclear non-proliferation experts, but also the situation of the US government's responsible agencies for nuclear non-proliferation and nuclear power policies. North Korea conducted its first nuclear test in 2006, which is after the 31st review meeting was held in 2004, but the situation at that time and the US thinking have not changed much, so it is introduced here (Fetter 2004).

The speaker said the following:

- Deregulation has been carried out, and cost differences are important in looking at the competitiveness of various energy forms.
- In the case of Japan, it is no problem to use plutonium in MOX fuel. There is a big concern about North Korea.
- At the nuclear non-proliferation meeting, when asked "why only Japan can reprocess" from countries such as South Korea, Taiwan, and Iran, it is hard to answer.
- It is possible to cause a nuclear explosion even with civilian plutonium.
- The United States decided on the once-through policy during the Ford administration in 1975, and this was continued under the Carter administration. This was reversed under the Reagan administration, but the power industry decided

that reprocessing was not economically attractive. The Clinton administration reaffirmed the once-through cycle and reaffirmed the policy of suppressing reprocessing worldwide, but countries like Japan, with a comprehensive nuclear non-proliferation mechanism and a proper reprocessing program, were exceptions. Under the Bush administration, the so-called fourth-generation nuclear reactor program, which opens the door to advancing the fuel cycle, was started. As far as I can see, the U.S. power industry is not interested in fast breeder reactors.
- The uranium market is very open, and competition has become more intense compared to the late 1960s.
- The direction of nuclear power is increasingly determined by market principles. Even if the Bush administration puts forward policies to promote FBR and reprocessing, it will likely have little impact in reality.
- U.S. government officials in charge of nuclear non-proliferation believe it would be desirable for reprocessing to disappear worldwide. When I talk to the person in charge of nuclear policy at the Department of Energy, they view Japan's reprocessing more favorably.

The person in charge of U.S. non-proliferation policy is not in the Department of Energy, but in the Department of State. There is also a non-proliferation officer in the White House. The Department of State is a powerful agency responsible for foreign affairs and other U.S. national interests. In Japan's nuclear industry, information about the budget of the Department of Energy is often conveyed, but Japan is under the U.S. nuclear umbrella and needs to cooperate with U.S. non-proliferation policy. Japan needs to consider measures for the issue of stable energy supply, regardless of reprocessing and fast breeder reactors.

References

Bunn M et al (2003) The economics of reprocessing vs. direct disposal of spent nuclear fuel, December 2003, Harvard University. https://dash.harvard.edu/bitstream/handle/1/29914177/Bunn_Economics-of-Reprocessing-vs-Direct-Disposal-of-Spent-Nuclear-Fuel.pdf?sequence=1&isAllowed=y. Accessed 12 Aug 2024

Fetter S (2004) The Economics of Reprocessing and Recycle v. Direct Disposal, 31st Fukushima Prefecture Energy Policy Review Meeting, Fukushima Prefecture Energy Division, May 31, 2004. https://www.pref.fukushima.lg.jp/uploaded/attachment/14537.pdf. Accessed 12 Aug 2024

Chapter 11
Liquid Fuel Reactors and Molten Salt Reactors

The liquid fuel reactors that have been built in the United States are aqueous homogeneous reactors and molten salt reactors. Both were built up until the 1960s. The fact that the fuel is liquid is a major difference from other types of reactors where solid nuclear fuel material is used with fuel cladding. Both aqueous homogeneous reactors and molten salt reactors are being considered for the possibility of thermal neutron breeding reactors in the thorium cycle. Liquid metal fuel reactors were also considered, but were not built due to the solubility of nuclear fuel material and fission products in liquid metal (bismuth) and the problem of material corrosion.

The aqueous homogeneous reactor is a nuclear reactor that uses salts of fissile materials dissolved in water as fuel, such as uranyl nitrate or uranyl chloride. It can reach criticality with natural uranium when using heavy water. In the United States, it was considered as a military plutonium production reactor during World War II, but due to the need for a large amount of heavy water, a natural uranium fuel graphite-moderated light water-cooled reactor was used for plutonium production. The principle of the aqueous homogeneous reactor was demonstrated during World War II. Enrico Fermi created research reactors (critical experiment devices) called LOPO, HYPO, and SUPO in sequence at the Los Alamos Laboratory in the United States. In research aimed at utilizing nuclear energy, aqueous homogeneous reactors, LAMPRE-1 (thermal output 2 MW) and LAMPRE-2 (thermal output 1.3 MW), which used enriched uranium as fuel, were built at the Los Alamos Laboratory in the 1950s. LAMPRE-1 operated for 5 h at a thermal output of 20 kW, but was stopped due to material corrosion. At the Oak Ridge Laboratory in the United States, aqueous homogeneous reactors HRE-1 (thermal output 2 MW) and HRE-2 (thermal output 5 MW) were built in the 1950s. HRE-1 has been operating for 24 months at a thermal output of over 1600 kW and a maximum temperature of 482F (250C). HRE-2, which used heavy water, operated until 1961 (Loftness 1964, Chap. 10).

The aqueous homogeneous reactor is also called a water boiler reactor. This name was given because the gas generated by radiation appears as bubbles in the aqueous solution. The aqueous homogeneous reactor is also used as a research reactor, which

Y. Oka, *Nuclear Power Reactor Development*,
https://doi.org/10.1007/978-981-96-2731-8_11

is sometimes called a water boiler reactor. SUPO in Los Alamos was used until 1974. In Japan as well, a critical experiment device of the aqueous homogeneous reactor was built in the 1960s at the then Japan Atomic Energy Research Institute (now the Japan Atomic Energy Agency). At that time, the use of thorium was also considered. In the 1990s, STACY and TRACY, aqueous homogeneous reactors, were built as mock-up experiment devices for critical accidents at reprocessing facilities. STACY is still in use today.

The molten salt reactor is a liquid fuel reactor that uses a mixture of fluoride of nuclear fuel material in the parent material, lithium and beryllium fluoride molten salt (FLiBe, LiF-BeF2), as fuel and heat transport medium. Molten salt is chemically stable, non-flammable or difficult to burn, has low vapor pressure at high temperatures, high radiation resistance, etc. Alkali metal halogen-based molten salts such as FLiBe, which easily dissolve strong ionic salts, are used in industrial processes such as chemical production by molten electrolysis. Molten salts with high specific heat and latent heat of fusion are used as heat storage materials and heat media for solar thermal power generation.

The first molten salt reactor made was the Aircraft Reactor Experiment (ARE); a type of aircraft nuclear reactor was operated at the Oak Ridge National Laboratory (ORNL) in the United States in 1954 with a thermal output of 2.5 MW and a maximum temperature of 1500F (816 C) for 460 h. NaF-BeF-UF4 was used as molten salt fuel. Next, the Molten Salt Reactor Experiment (MSRE) was built at ORNL and became critical in 1965, operating until 1969. The thermal output is 7.4 MW. The MSRE is graphite-moderated, and the inlet and outlet temperatures of the primary molten salt (melting point 434 C) are 635 C and 663 C, respectively. The heat of the primary molten salt is exchanged with the heat medium of the intermediate loop using an intermediate heat exchanger, and it is cooled by an air-cooled heat exchanger. This cooling system configuration is similar to that of a loop-type fast breeder experimental reactor (for example, JOYO). The MSRE was initially operated with Uranium-235 fuel, and from 1968 with Uranium-233.

The configuration of the cooling system of the concept considered as a molten salt reactor for power generation (Yoshioka 2013, Fig. 1.1) is similar to that of a loop-type sodium-cooled fast reactor (Fig. 1.1). The similarities between the molten salt fuel power reactor and the sodium-cooled fast reactor include the need for an intermediate heat exchanger (primary heat exchanger) and a secondary cooling system that transfers heat to the steam generator, the liquid level of the primary coolant is at the top of the reactor vessel, and the melting point of the molten salt is higher than room temperature (450 C in FLiBe), so it is necessary to keep the piping heated above the melting point even when stopped.

A unique challenge for molten salt reactors is that nuclear fuel materials containing fission products circulate in the primary cooling system, so as was the case with the MSRE, the primary cooling system, molten salt drain tank, and exhaust system need to be installed in a hot cell and maintained by remote operation (Blumberg 1968). In the case of sodium-cooled fast reactors, the sodium becomes radioactive, but with a half-life of about 15 h, after a few days, humans can enter the site and perform maintenance work on valves, pumps, and measuring instruments in the primary cooling system. In the case of light water reactors, the half-life of radioactive nitrogen-16 produced by the radio activation of oxygen, a constituent element of water, is 7 s, so there is no need to wait for its decay before performing maintenance work after stopping the reactor. The primary cooling system of a molten salt reactor, even after stopping the reactor and draining the molten salt into a drain tank, cannot avoid the slight adhesion and residual of molten salt containing nuclear fuel material and fission products in the primary cooling system piping, resulting in extremely high radiation levels. Therefore, maintenance must be performed using a hot cell (a room surrounded by thick concrete shielding with thick glass windows). Work inside the hot cell requires the use of a magic hand, and fine repair work that can be done by hand is not possible, so the main task is to replace equipment. When removing equipment from the hot cell, it must be decontaminated. The image of a molten salt reactor is not like that of a light water reactor or a sodium-cooled fast reactor, but is closer to the image of a reprocessing plant with equipment such as a reactor, intermediate heat exchangers, and molten salt drain tanks inside the hot cell. This can be understood by looking at photos of the MSRE (Blumberg 1968, Fig. 1.2). A report examining the maintenance methods of the Oak Ridge National Laboratory (ORNL) molten salt breeder reactor (Blumberg 1967) states as a policy to use as reliable equipment as possible and replace it if it fails. Compared to light water reactors, molten salt reactors are thought to pose challenges not only in terms of capital costs, but also in terms of maintenance costs and operational rates.

The concept of a molten salt reactor proposed by Yoshioka and others is a thermal neutron reactor using graphite as a moderator, similar to the MSRE of ORNL, but the concept of a molten salt fast reactor without graphite has also been proposed in the U.S. and other countries. When using graphite as a moderator, as mentioned in the section on graphite-moderated reactors, the power density is lower than that of light water reactors, and the reactor core becomes larger, resulting in higher construction costs. The configuration of the cooling system of a molten salt reactor is similar to that of a sodium-cooled fast reactor. As already mentioned, the construction cost of a pool-type sodium-cooled fast reactor (Superphenix, 1200 MWe) is reported in the 2007 IAEA report (IAEA 2007, page 84, Table 7). It is 2.66 times that of France's 1400 MWe pressurized water light water reactor. The molten salt power reactor has a cooling system configuration similar to that of a loop type fast reactor, and the construction of the loop-type fast reactor is slightly higher than that of the pool type. In the case of a molten salt reactor, the primary cooling system flows molten salt containing fuel and fission products, so it is necessary to enclose it in a hot cell, and the construction cost is expected to be substantially higher than that of a loop-type sodium-cooled fast reactor.

References

Blumberg R (1967) Maintenance development for molten-salt breeder reactor. ORNL-TM-1859. Oak ridge national laboratory. https://www.osti.gov/servlets/purl/4355531. Accessed 12 Aug 2024

Blumberg R, Hise EC (1968) MSRE design and operations report part X, maintenance equipment and procedures. ORNL-TM-910. https://www.osti.gov/servlets/purl/4527944. Accessed 12 Aug 2024

IAEA (2007) Liquid metal cooled reactors: experience in design and operation. IAEA-TECDOC-1569. IAEA. https://www-pub.iaea.org/MTCD/Publications/PDF/te_1569_web.pdf. Accessed 12 Aug 2024

Loftness RL (1964) Nuclear power plants. Chapter 10 Fluid fuel Reactors, Van Nostrand

Yoshioka R, Kinoshita M (2013) Current status of thorium molten salt reactor development. 17th Nuclear Committee, Document no. 2–2 Japan Atomic Energy Commission (in Japanese) https://www.aec.go.jp/kaigi/teirei/2013/siryo17/2-2_haifu.pdf. Accessed 4 April 2025

Chapter 12
Organic Cooled Reactors

Organic materials such as diphenyl or oxidized diphenyl (diphenyl ether) have been used as high-temperature heat transport media in the chemical industry since the 1930s. The organic material-cooled reactor was built in the United States from the 1950s to the mid-1960s, with the experimental reactor OMRE (thermal output 16 MW, operation 1957–63) at the Idaho National Laboratory, and Piqua (thermal output 45MW, electric output 12.5 MW, coolant inlet/outlet temperature 270C/301C, primary system pressure 724 kPa, operation 1963–1966) was built in Ohio (Loftness 1964). The Piqua power plant was built by the local public power company and the U.S. government in accordance with the U.S. government's small reactor construction plan at the time. The coolant of these reactors is the aromatic compound terphenyl. In addition, there are the Soviet ARBUS (thermal output 2 MW) and the Canadian heavy water-moderated organic material-cooled reactor WR-1 (thermal output 30 MW, operated as an organic material-cooled reactor from 1965 to 1973, and then operated with heavy water cooling until 1985).

Organic liquids generally have the advantage of being less corrosive to metals than water, but their specific heat is smaller than water, and their cooling performance is inferior to water. At Piqua, finned fuel cladding tubes are used to promote heat transfer. The biggest problem is the decomposition of the organic coolant by radiation, which generates hydrogen and produces short-chain bonds and long-chain organics. Long-chain molecules adhere to the fuel cladding tube and hinder continuous operation. Care must also be taken with flammability and toxicity. The organic material-cooled reactor operated only for a short period of time.

After the nuclear fuel was removed from the Piqua power plant, the containment vessel was left in place, the area around the reactor vessel was filled with concrete, the floor above was covered with a waterproof sheet, and the entombment disposal was carried out. The radiation levels of sump water at the bottom of the containment vessel and other areas are being monitored by the federal government's Center for Disease Control and Prevention (CDC). The power plant buildings other than the

Y. Oka, *Nuclear Power Reactor Development*,
https://doi.org/10.1007/978-981-96-2731-8_12

containment vessel have been decontaminated and are being used as warehouses and office space for the city of Piqua.

Reference

Loftness RL (1964) Nuclear Power Plants. Chapter 6 Organic cooled reactors, Van Nostrand

Chapter 13
Nuclear Reactors and Radioisotope Power Sources for Aerospace

Nuclear power is used in space exploration where sunlight does not reach and in space missions that require large amounts of power. There has been development of nuclear reactors for aerospace use (reactors that generate thrust using nuclear fission energy). This is not well known, so the author will also discuss ground tests of reactors that were not commercialized.

The development of nuclear reactors for aviation use (reactors that generate jet engine thrust using nuclear fission energy) was carried out in the United States from the 1950s to the early 1960s, but it did not reach practical use. Tests were also conducted to generate rocket thrust after launch with rocket fuel using a nuclear reactor, but they did not reach practical use.

Nuclear reactors as power sources for space were commercialized in the 1960s and used more than 30 times as power sources for artificial satellites in the former Soviet Union. In the United States, it was used in SNAP10 in the 1960s. The Radioisotope Thermoelectric Generator (RTG), which converts the heat generated by radioisotopes (such as plutonium 238) into thermal electricity, has been commercialized and has been used as an important power source in space since 1961. It is based on the Seebeck effect. It is a phenomenon in which a temperature difference between two dissimilar electrical conductors or semiconductors produces a voltage. A method of converting thermal electrons to electricity (thermionic conversion) was also developed in the former Soviet Union and used once in space. They are explained in the following.

The development of nuclear reactors for aerospace use began in the United States after World War II as the Advanced Nuclear Propulsion (ANP) program. Initially, air-cooled reactors and those that used liquid metal cooling and intermediate heat exchangers to transfer nuclear heat to air were considered. Air-cooled reactors were adopted and the experimental reactors Heat Transfer Reactor Experiment (HTRE)-1, 2, 3 were tested at the Idaho Nuclear Reactor Test Site from 1955 to 1958. HTRE-1 and 2 were water-moderated, and the temperature and pressure of the cooling air were 1335 F (723 C) and 55psig (379 kPa), with a thermal output of 17.5 MW. HTRE-3 was zirconium hydride-moderated at 1435 F (779 C), 54psig (372 kPa), 32.4 MW.

Y. Oka, *Nuclear Power Reactor Development*,
https://doi.org/10.1007/978-981-96-2731-8_13

HTRE-3 was tested by connecting to two turbojet engines. The ANP program ended in 1961, and as for the PLUTO program, the air-cooled TORY-II (thermal output 155 MW, cooling air temperature 1975 F (1079 C), pressure 350psig (2.41 MPa)) with beryllium oxide moderation was tested in Nevada in the early 1960s. The type of jet engine was turbojet for HTRE and ramjet for TORY. The design and testing were conducted by GE.

For use in jet engines, a turbine inlet temperature of 1850 F (1010 C) is required, which requires a temperature of 2300 F (1260 C) in the nuclear reactor. However, there were no metal materials for cladding tubes and structural materials that could withstand this. The fuel was a mixture of highly enriched uranium oxide and nickel-chromium alloy (80 Ni/20 Cr), and the fuel cladding tube also used nickel-chromium alloy (80 Ni/20 Cr). The design life of the HTRE was 1000 hours. Another problem was that there were restrictions on the size and weight of the nuclear reactor for aircraft. The design life required for practical use was short. The requirements for radiation shielding differ between manned and unmanned aircraft. It seems that unmanned aircraft were assumed in TORY.

The development of using a nuclear reactor for rocket propulsion was undertaken by the Los Alamos National Laboratory as the ROVER program, and was conducted at the Nevada Test Site from 1955 to 1973. It was started as a project of the U.S. Air Force, but after the shock of the Soviet Union's satellite launch, Sputnik, NASA took over from 1958 and it was carried out as part of NASA's Nuclear Engine for Rocket Vehicle Application (NERVA) project. The ROVER's nuclear reactors were KIWI (1955–1966), Phoebus (1964–1969), and Pewee (1969–1972), with KIWI and Phoebus being large nuclear reactors, and Pewee being a small nuclear reactor due to budget constraints. KIWI is named after a flightless bird from New Zealand. The structure of a nuclear-powered rocket is simple, heating hydrogen in a nuclear reactor and ejecting it from a rocket nozzle to obtain thrust.

Highly enriched uranium was used as fuel for these nuclear reactors, graphite was used as a moderator, and hydrogen was used as a coolant and rocket propellant. Nuclear propulsion was expected to be used not in the first stage of the rocket, but in the later stages. When using a nuclear reactor, compared to chemical rockets, the persistence of output is superior. Furthermore, it is easy to repeat start and stop, and by utilizing its load-following capability, it is also possible to easily change the output, making it more efficient (specific impulse) than chemical rockets. The fact that the propellant hydrogen is light was also an advantage. The control of the output of the nuclear reactor was done by rotating a cylinder coated with the neutron-absorbing material boron around the nuclear reactor. The increase in thrust could be achieved not by rotating the cylinder, but by increasing the amount of hydrogen ejected. By increasing the amount of hydrogen supplied, the number of neutrons slowed down increased, the number of nuclear fissions increased, the output of the nuclear reactor increased, and the thrust also increased.

The challenge with nuclear rockets is that, in chemical propulsion rockets, the temperature of the propellant is 6000 F (3316 C), higher than the temperature of the combustion chamber wall, 1400 F (760 C), but in nuclear rockets, conversely, the temperature of the heat transfer surface of the nuclear fuel elements needs to be higher

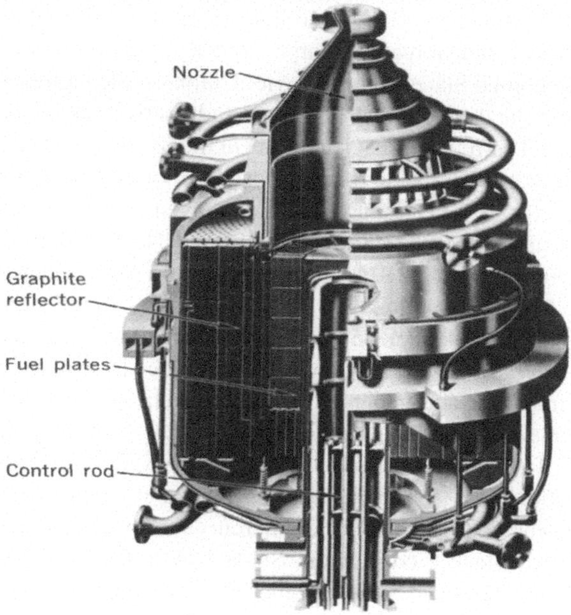

Fig. 13.1 Nuclear rocket propulsion engine (KIWI). *Source* Atomic energy commission, USA

than the temperature of the propellant. When hydrogen was used as a propellant, the temperature of the combustion chamber could be 3000 F (1649 C), and it could perform the same as a chemical rocket using a 5000 F (2760 C) oxygen-hydrogen compound as a propellant, but the weight of the nuclear reactor and the complex control system became drawbacks. Graphite-based materials were considered the best in terms of high-temperature strength, hydrogen corrosion resistance, small neutron absorption rate, and thermal stress and thermal conductivity. The cross-section of the nuclear rocket engine KIWI is shown in Fig. 13.1. The fuel elements are made of U235 evenly distributed on flat graphite plates, which are arranged inside the graphite reflector at 1.27 mm intervals to form a cylindrical core. The cooling hydrogen descended the reflector, flowed upward from the bottom of the core through the fuel elements, and was ejected from the nozzle. The KIWI-A test was conducted in 1959, and the nuclear reactor was operating at rated output for several minutes. The operating temperature of 3000 F (1649C) was achieved in the KIWI-A' test conducted in 1960. KIWI was tested eight times by 1964. After that, Phoebus was tested three times and Pewee once in the 1960s. Phoebus 2 A achieved an average total output of 4082MW, a total output time of 744 seconds, and a propellant exit temperature of 2283K. NASA ended the NERVA program, including ROVER, in 1973. The nuclear propulsion rocket did not actually fly.

For power supplies exceeding several hundred watts, power sources using a nuclear reactor as a heat source become advantageous over solar power and radioactive isotope thermoelectric power sources. In deep space exploration where sunlight does not reach, a power source other than solar power generation is required. Fuel cells require fossil fuels, so long-term use is difficult. Research and development to make a nuclear reactor a large-capacity space power source began in the late 1950s in the United States and the Soviet Union. In the United States, it was conducted along with the development of a space radioactive isotope power source (RTG, Radioisotope Thermoelectric Generator) in the System for Nuclear Auxiliary Power (SNAP) program. The name SNAP is used in both of these plans, but those using a nuclear reactor as a heat source are given even numbers, and there are SNAP-2 (thermal output 3 kW, operation 1961–1963), SNAP-8 (thermal output 1000 kW, operation 1969), and SNAP10A (thermal output 45.5 kW, electrical output 650 W, launched in 1965) in the 1960s. Also, the SP100 of the 1980s is a nuclear heat source. SNAP2 and SNAP8 use a nuclear reactor with NaK liquid metal coolant and zirconium hydride moderator as a heat source, and transfer heat to mercury in an intermediate heat exchanger, and generate electricity with a turbine driven by mercury vapor. After rotating the turbine, the mercury vapor is cooled by a radiator to space.

The first SNAP nuclear reactor, SNAP2, operated on the ground from 1959 to 1960. SNAP10A, while having the same reactor part as SNAP2, did not generate power through a secondary mercury turbine, but generated 300W of power through thermoelectric conversion. SNAP10A is the only satellite launched in the United States equipped with a nuclear reactor heat source. Prior to the development of SNAP2, an experimental reactor called SNAP Experimental Reactor (SER, thermal output 50 kW, operated 1959–1960), which is cooled by NaK and moderated by zirconium hydride, was created. As the demand for large-capacity space power sources waned, the focus shifted in the early 1970s from space power sources using nuclear reactors to RTGs using radioisotopes (Pu238) as heat sources. Subsequently, from 1983 to 1994, there was research on SP100, which uses a nuclear reactor as a heat source and conducts thermoelectric power generation by transmitting heat with a heat pipe.

The Soviet Union, from 1967 to 1988, sent a nuclear power source to space 31 times, which uses a nuclear reactor called BES-5 as a heat source and generates power through thermoelectric conversion. Furthermore, from 1960 to 1970, they developed the TOPAZ nuclear reactor that uses thermionic conversion. The first time it was used in space was in 1987. After the end of the Cold War, the United States attempted to use TOPAZ and conducted ground tests, but it has not been used in space. In 2008, NASA announced that it would test the elemental technology for using a small nuclear reactor on the moon or Mars. Both SNAP10A and the Soviet Union's BES-5, the nuclear reactor power sources used in space, generate power through simple thermoelectric conversion, not the Rankine cycle that requires heat exchangers and turbines. Furthermore, the nuclear reactor is launched into orbit and then started. When the satellite's life ends, it can be prevented from falling back to Earth.

Radioisotope Thermoelectric Generators (RTGs) that convert the decay heat of radioactive isotopes into electricity have been used not only for space probes in the weak sunlight of space, but also for power sources such as communication equipment in the polar regions. The radioactive isotopes used are Plutonium-238 (half-life 87.7 years, power density 0.54 W/g) and Strontium-90 (half-life 28.8 years, power density 0.46 W/g). Plutonium-238, which mainly undergoes alpha decay, is easy to shield from radiation, but requires special equipment to produce. The RTGs for space use in the United States use Plutonium-238. Large RTGs (thermal output 4.4 kw, electrical output 300 W) were used for space exploration by Galileo (Jupiter probe, launched in 1989) and Cassini (Saturn probe, launched in 1997). RTGs with a thermal output of 1.48 kW and an electrical output of 73 W were used from Apollo 12 to 17 for lunar exploration. Those using Strontium-90 as a heat source require radiation shielding, but are relatively easy to produce, so they are used in Soviet space RTGs. RTGs are used not only to supply power to spacecraft communication equipment, but also to maintain the operating temperature of measuring instruments and machinery. The United States also uses strontium compounds as a heat source for polar communication equipment. When the equipment supplied by the polar RTG has completed its role, it is necessary to recover the radiation source. References for this chapter include, for example, Chapter 11 of (Loftness 1964; DOE 2015).

References

DOE (2015) Atomic Power in Space II. Department of Energy, INL/EXT-15-34409. https://inldigitallibrary.inl.gov/sites/sti/sti/Sort_63995.pdf. Accessed 12 Aug 2024

Loftness RL (1964) Nuclear power plants. Chapter 11 Aerospace reactors, Van Nostrand

Chapter 14
Generation III and Generation IV Nuclear Reactors

The Generation III reactor is an improved light water reactor and improved CANDU based on previously built commercial power plants. Generation IV reactors are six types of advanced reactors which were selected by the Generation IV international forum in 2002.

The Generation IV International Forum (GIF) was initiated by the U.S. Department of Energy in 2000 and officially launched in 2001 with the participation of nine countries (Argentina, Brazil, Canada, France, Japan, South Korea, South Africa, the UK, and the USA) that use nuclear power generation. After its establishment, four countries (Switzerland, China, Russia, and Australia) and the European Union (EU) have joined. The purpose of the GIF is not to build reactors, but to share research and development information. After about two years of discussion, out of about 100 reactor concepts, six types of reactors were selected as Generation IV reactors (WNA 2020). These are the Gas-cooled Fast Reactor (GFR), Lead-cooled Fast Reactor (LFR), Molten Salt Reactor, Sodium-cooled Fast Reactor (SFR), Supercritical Water-Cooled Reactor (SCWR), and Very High-Temperature gas-cooled reactor (VHTR). Among these, all except the supercritical pressure light water-cooled reactor have had similar reactor concepts researched or developed in the past, and the configuration of their reactor lineage is the same as, similar to, or a combination of the technologies of the high-temperature gas-cooled reactor, molten salt reactor, and liquid metal-cooled fast reactor which are already mentioned in other chapters. The gas-cooled fast reactor has been considered in Europe in the past as an alternative to the sodium-cooled fast reactor. The GIF Secretariat has been handled by the OECD/NEA since 2005, and their homepage contains explanations of these reactor concepts. The Generation IV reactor was initially considered a reactor concept aiming for practical use from 2010 to 2030.

The Generation III reactor is an improved light water reactor and improved CANDU based on previously built commercial power plants. The U.S. Department of Energy launched the NP2010 (Nuclear Power 2010) plan in 2002, aiming to start commercial reactor operation in 2010. In partnership with private companies, the

Y. Oka, *Nuclear Power Reactor Development*,
https://doi.org/10.1007/978-981-96-2731-8_14

goal was not only to develop improved light water reactor technology, but also to find construction sites and obtain permits. The approval for new construction of nuclear power plants in the United States was improved so that the review of individual plants was limited to site-related issues, and the construction and operation were reviewed together, making it possible to streamline and shorten the review process. That is, it is necessary to obtain Design Certification (DC) and Combined Construction and Operating License in order. In NP2010, if the industry bears more than 50% of the cost for obtaining these permits (half each for the reactor manufacturing company and the power company), government support can be obtained (Johnson 2002).

Furthermore, the U.S. introduced a system in the 2005 Energy Policy Act that guarantees the debt of construction plans for renewable energy and nuclear power plants using innovative and clean technology. As a result, many new movements toward the construction of new light water reactors started in the United States in the 2000s. The construction of the AP1000 was at the forefront. A Japanese reactor manufacturing company acquired WH, a U.S. reactor manufacturing company, from BNFL in the UK and entered the U.S. market. Another Japanese reactor manufacturing company also placed a nuclear power expert at the top of its U.S. corporation and tried to sell the US-APWR to the United States (Yamauchi 2019).

However, subsequently, due to the construction delay of the leading AP1000 and the emergence of shale gas and the subsequent decline in the relative economic advantage of nuclear power in the United States due to the decline in natural gas prices (Yamauchi 2019, p 17), new construction in the United States was significantly reduced, and only the construction of the AP1000 at Vogle 3 and 4, which was ordered by the power company, continued. WH went bankrupt due to the delay of the construction, and its parent company, a Japanese company, was also hit hard.

Since the late 2000s, in the United States and elsewhere, there has been an increase in activities aimed at the new construction of small reactors (nuclear reactors with an electrical output of 300MW or less) by venture companies. The government is also supporting these activities through cost-sharing programs and the like. Information about the reactors these venture companies are advancing is only minimally disclosed. This book mainly discusses commercial power reactors that have been built and operated, so it does not touch on these activities. The situation of small power reactors after 2010 can be found on the homepage of the World Nuclear Association and elsewhere (WNA 2024c). In addition, activities are being carried out for the construction of large light water reactors, not only enhanced light water reactors with passive safety facilities, but also light water reactors that are an extension of the light water reactors developed and built by nuclear reactor manufacturers and companies in various countries.

Venture companies aiming for the commercialization of power reactors need not only experts in nuclear reactor technology, but also experts in strategy, finance, regulation and licensing, supply chain, human resources, etc. In terms of nuclear reactor technology, in addition to experts in fuel materials, reactor and plant design, analysis, testing, manufacturing, etc. experts in technologies unique to that reactor are also needed. This can be understood by looking at the organization of small power reactor venture companies in the United States. In the case of existing light water reactor manufacturers, these human resources are secured within the company.

While the information on the reactors being developed by companies is only minimally disclosed, information on research and development activities is widely disclosed by research institutions and researchers. International activities like GIF are also being carried out. This information is easily disseminated through newsletters of nuclear-related organizations and societies. Research institutions and researchers are researching because they think or like the reactor concept. They also want to secure research funding. Unilateral information may be disseminated and spread for the purpose of obtaining research and development funds. The more specialized the field, the fewer the number of experts in that field. When investing national research funds, care must be taken as the information circulating in the public is not always neutral.

Some countries have established and operate a check-and-review mechanism for the results of administrative work, but in many countries, the reflection of past experience in the selection of research plans related to the practical application of research and development is insufficient. Power company officials are not experts in research and development, and even when there is an advisory committee, it tends to be influenced by the opinions of groups that have been involved in research and development in the past. A few university professors have experience in building nuclear reactors, and they are not necessarily neutral because they receive support such as research funds from the responsible ministries. The field of nuclear researchers is subdivided into reactor physics, heat transfer fluid, materials, etc. and there are very few nuclear researchers who have a comprehensive knowledge of the history and technological changes of nuclear reactors, construction investment, etc. As a result, it cannot be said that the experience of past commercial nuclear reactor research and development is sufficiently reflected in the planning and theme selection of research and development. There is no document that summarizes past experiences related to the use of nuclear power generation from a neutral standpoint based on expert knowledge.

Furthermore, in order to advance the research and development in charge, there may be a so-called "argument for the sake of" by research and development institutions and researchers. It should be noted that this can be a hindrance to advancing the use of nuclear power. For example, in the case of geological disposal, the argument that the burden of geological disposal can be reduced by reducing the long half-life transuranic elements by incineration treatment using a fast reactor is issued from the researchers of the fast reactor and its research and development institution. However, the exposure risk of geological disposal is due to water-soluble fission products. Transuranic elements are not water-soluble and do not contribute to the exposure risk of geological disposal, so this opinion is off the mark. This argument is due to the fact that nuclear reactor experts do not know about geological disposal, and it is due to their "argument for the sake of" to secure their research budget. For example, the French Nuclear Regulatory Commission has stated that "Separation and transmutation cannot dispose of all high-level radioactive waste, long-term storage is not a permanent solution, deep geological disposal is the only solution" (Andra 2014).

The dissemination of safety and risk information to the public should be left to the nuclear regulatory agency, which is responsible for ensuring the safety of the public. Even if it does not go as far as an argument for the sake of, if you do not know or ignore past experiences, similar failures will be repeated. In this book, we have described the experience gained from commercialized power reactors in terms of research and development based on facts. In future research and development, it is necessary to face the experience of past research and development and commercial use and consider new measures based on it. Breakthroughs should come from learning from past experiences and facing problems head-on.

References

Andra (2024) More than 25 years research. International Andra. Accessed 26 Apr 2024. https://international.andra.fr/projects/cigeo/more-25-years-research. Accessed 12 Aug 2024

Johnson S (2002) Nuclear Power 2010, Program overview, presentation to the nuclear energy research advisory committee. Office of nuclear energy science and technology, department of energy, April 15, 2002. https://www.energy.gov/sites/prod/files/Presentation%20-%202010%20Program%20Overview%20-%20Presentation%20to%20the%20NEAC.pdf. Accessed 12 Aug 2024

WNA (2020) Generation IV nuclear reactors (updated December 2020). World Nuclear Association. https://world-nuclear.org/information-library/nuclear-fuel-cycle/nuclear-power-reactors/generation-iv-nuclear-reactors.aspx. Accessed 12 Aug 2024

WNA (2024c) Small nuclear power reactors (updated February 2024). World Nuclear Association. https://www.world-nuclear.org/information-library/nuclear-fuel-cycle/nuclear-power-reactors/small-nuclear-power-reactors.aspx. Accessed 12 Aug 2024

Yamauchi K (2019) Current status and challenges of the U.S. Nuclear industry licensing: perspectives from business experience in the U.S., Atomic energy society of japan nuclear safety subcommittee. In: 7th summer seminar, August 19, 2019 (in Japanese). http://www.aesj.or.jp/~safety/pdf/summerseminar/04_yamauchi_20190819.pdf. Accessed 12 Aug 2024

Chapter 15
Investment Risk Reduction and Power Generation Cost Reduction

Nuclear power plants, along with large-scale hydroelectric power plants, are the largest structures ever built by mankind, and their construction takes a great deal of money and time. They, however, produce low cost electricity after amortizing the initial investment. In the development and practical application of nuclear reactors, issues such as the investment risk of new construction and power generation costs are more important than the technical aspects. This chapter explains the investment risks associated with building new nuclear power plants and reducing power generation costs. These issues are discussed in detail in the sister book "Nuclear Power and Society, Sects. 1.6 and 1.7".

When constructing a new nuclear power plant with private investment, there are unique investment risks not only in technology but also in design, manufacturing, construction, social, political, and regulatory aspects compared to other power generation methods. Nuclear power plants have a longer construction period, larger initial investment, and greater investment risk compared to thermal power plants. This greatly affects the profitability of the construction project, making it a major concern for both lenders and borrowers of construction funds. In the construction of new nuclear power plants, it is essential to deal with this investment risk. The power plant construction side seeks to share this investment risk with project participants through contracts, but usually needs to pay a premium associated with the risk to the risk bearers. From the investor's perspective, profits from the investment must be obtained in proportion to the size of the investment risk. Therefore, the investment structure is affected by the investment risk. The investment structure affects the interest rate on borrowed money. In other words, if the investment risk is high, the interest rate on borrowed money will be high. If the construction period is long, the proportion of interest in the construction cost becomes large. For example, if the construction period is 5 years, the interest rate on borrowed money may be 5%, but if the construction period is 7 years, it may be 10%, in which case the amount of borrowed money payment is about three times more in the latter case (IAEA 2017, Fig. 15.3).

Y. Oka, *Nuclear Power Reactor Development*,
https://doi.org/10.1007/978-981-96-2731-8_15

Power plants do not generate profits until they start generating power. Since the construction cost of the power plant is covered by borrowed money, the shorter the construction period, the better. Delays in construction make it difficult to repay borrowed money, increasing debt and becoming a major investment risk. Those who invest in construction, such as power operators, recover their investments through profits after the start of power generation, so the power generation cost needs to be smaller compared to other power generation methods. Therefore, in the construction of new power plants, it is necessary to reduce both investment risk and power generation cost.

New technology can be a cause of investment risk. For example, the AP1000 adopted a new technology called a canned motor pump for the primary coolant pump, but struggled to meet the safety regulation standards for flow rate reduction when the pump stopped, becoming one of the causes of construction delay. Instrumentation and control technology is advancing rapidly, and there have been cases where delays in response, including standard criteria, have been a factor in construction delays in past multiple light water reactor construction projects. Delays in licensing approval for new technology can be a cause of delay.

From the perspective of reducing investment risk, it is concluded that constructing large light water reactors that were built in the past without changing the design has fewer uncertainties and can minimize investment risk. In this case as well, if construction has not been carried out for many years, there are investment risks such as the manufacture and supply of parts (supply chain), nuclear-specific construction work, such as seismic foundation work, and the experience of obtaining permits and approvals. In future nuclear power plant construction, it is necessary to respond not only to factors due to the design of the reactor, but also to these investment risk factors. This is a common issue not only for the construction of large light water reactors, but also for the construction of small modular reactors. The continuation of light water reactor construction reduces the risk of construction delay.

In the United States, there are nuclear power plants that have obtained operating permits for 80 years from the start of commercial operation. Like hydroelectric power plants, nuclear power plants can be used for a long time once built. The above explained about the construction of nuclear power plants by private finance, but there is a view that the development of infrastructure for stable power supply such as nuclear power plants, hydroelectric power plants, and transmission lines, like road construction, should be carried out with the involvement of the state because it cannot be covered by private finance. The investment environment for nuclear power plants varies by country, but it is noteworthy that in the UK, where economic liberalization has progressed, new investment models for nuclear power plants, such as the RAB model, are being explored.

In the United States, due to low natural gas prices, the competitiveness of light water reactors that have been generating power for many years is an issue, especially in deregulated states. When a large amount of renewable energy is introduced, the wholesale electricity rate may become zero in temporary competitive bidding when the weather is suitable for its generation, reducing the cost competitiveness of nuclear and thermal power generation. However, since renewable energy cannot always generate power, if nuclear power plants and thermal power plants are abolished, stable power supply and inexpensive power supply in the long term will be compromised.

Nuclear power generation reduces construction and operation management costs per kilowatt as the size (electrical output) of the power plant increases. Historically, the improvement in the economics of light water reactors has been achieved mainly by increasing power output. However, this approach is reaching its limit. Due to the power demand in the supply area of the power company and the constraints of the transmission network, the locations where power plants with huge single-unit electrical output can be built are limited. When a single power company builds and operates a new power plant, it is said that about one-tenth of its power supply is a guideline for the electrical output of the power plant to be built. For nuclear reactor manufacturers, it is said that a new design nuclear power plant needs to be built at least about 10 units to recover its design investment, so large-output power plants are disadvantageous in terms of design construction proficiency and inheritance of construction experience. The pursuit and practical application of technology and strategies for reducing power generation costs and the inheritance of construction experience are challenges for future nuclear power generation, along with reducing investment risk.

Venture companies are making various attempts, such as small modular reactors, mainly in the United States. The test reactors of these new types of reactors are small in scale and the investment amount is not as large as that of large commercial reactors. However, many of these new types of reactors, including non-light water reactors, are either the type of nuclear reactor mentioned earlier or a combination of their technical elements. If we expect further development of the test reactor, we need to learn from past experience. It can be considered that venture companies will cover investment risks with large founder profits when their business is successful and their stocks are listed.

Reference

IAEA (2017) Managing the Financial Risk Associated with the Financing of New Nuclear Power Plant Projects. IAEA Nuclear Energy Series, No NG-T-4.6. https://www-pub.iaea.org/MTCD/Publications/PDF/P1765_web.pdf. Accessed 12 Aug 2024

Chapter 16
Concept of Supercritical Pressure Light Water Reactors

The supercritical pressure light water reactor or Supercritical Water Reactor (SCWR) is a Generation IV reactor. It adopts once-through coolant cycle as supercritical coal fired power plants operating at supercritical-pressure. It is not a reactor that has been built in the past, but a reactor concept under research and development. The author noticed the advantage of simplifying light water reactors by using supercritical water for reactor cooling, and studied it at the university for 25 years from 1989. It was selected as one of Generation IV reactors in 2002. The author led the research group which created calculation codes and studied core design, startup and stability, safety, etc. It covers most of the design and analysis of power reactors and is explained in the chapter. It will be useful for understanding its outline. We will also introduce SCWR research in Japan and around the world.

Water, when pressurized and heated above 22.1 MPa (218 atmospheres), becomes a high-temperature fluid without boiling. Power generation by driving a steam turbine with this high-temperature fluid is a supercritical pressure light water reactor. Since there is no need for steam-water separation at supercritical pressure, the reactor cooling system becomes simpler than a light water reactor. The enthalpy per volume of "supercritical steam" is larger than that of subcritical pressure steam, making steam turbines and main steam pipes more compact. This reduces capital costs. Since there are no boiling point constraints, thermal efficiency also improves. Light water reactors were developed based on the subcritical pressure thermal power generation technology of the time, but thermal power generation has since transitioned to supercritical pressure. The experience of supercritical pressure thermal power generation can be used in the design, manufacture, operation, and maintenance of supercritical equipment. Since the supercritical pressure light water reactor is a once-through-flow type, like supercritical pressure thermal power plants, all of the reactor cooling water can be purified after it has become low-temperature condensed water. This is expected to significantly reduce the stress corrosion cracking problem that has plagued light water reactors.

Y. Oka, *Nuclear Power Reactor Development*,
https://doi.org/10.1007/978-981-96-2731-8_16

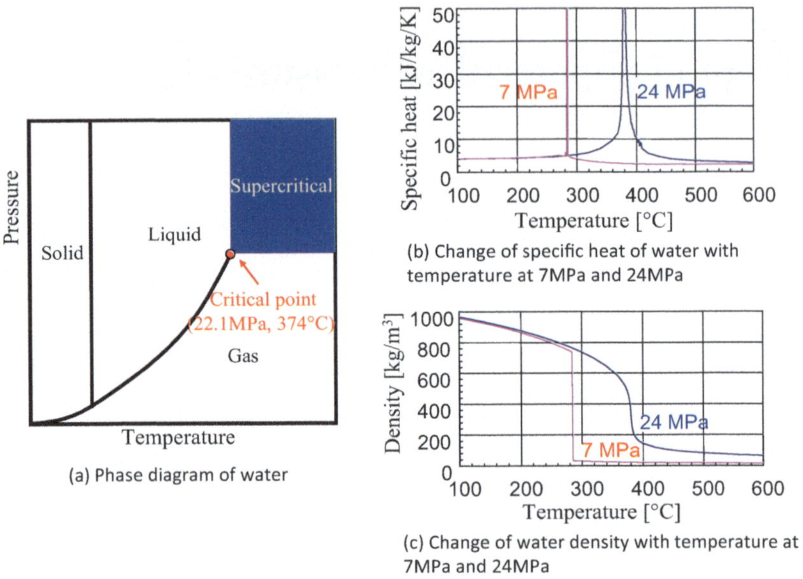

Fig. 16.1 Phase diagram and comparison of specific heat and density of water at 7 MPa and 24 MPa

Figure 16.1 shows the state diagram of water, and the changes in specific heat and density of water at supercritical pressure (24 MPa) and subcritical pressure (7 MPa). Water is solid ice at low temperatures below zero. When water is heated at pressures below 22.1 MPa, it boils and steam is generated from the water. However, when water is heated at pressures above 22.1 MPa, the boiling phenomenon disappears and it continuously changes from a high-density state (water) to a low-density fluid. This low-density fluid is conventionally called supercritical "steam". The 22.1 MPa is called the critical pressure of water. The corresponding temperature of water (critical temperature) is 374 C. These are called the critical points of water. Each pure substance of other liquids also has its own critical point. The critical point is the upper limit of temperature and pressure where phase transition between gas and liquid can occur. Under pressures higher than the critical point, the liquid does not boil even when heated, and its density decreases continuously. When water is heated above supercritical pressure, the boiling phenomenon disappears, but near the critical temperature (more precisely, the pseudocritical temperature), the density of water drops significantly.

When water is heated from a low-temperature state, at subcritical pressure, the density of water changes discontinuously at the boiling point (100 C at 1 atmosphere, 286 C at 7 MPa), and low-density steam (100 C at 1 atmosphere, 286 C at 7 MPa) is generated. The specific heat peaks at the boiling point allow for good heat removal. When heated at supercritical pressure, the density of water changes continuously, but as shown in the figure, it drops significantly around 380 C at 24 MPa. The point where

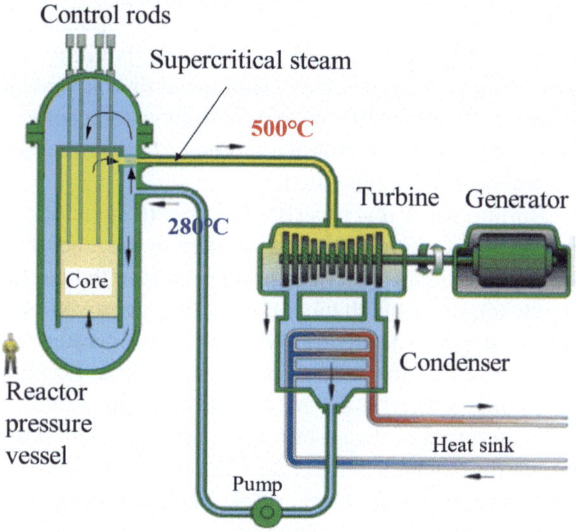

Fig. 16.2 Concept of SCWR (supercritical-pressure, light water cooled reactor). *Source* GIF 2023

the density drops significantly at supercritical pressure is called the pseudocritical point, and its temperature is called the pseudocritical temperature. The specific heat peaks at the pseudocritical point, allowing for good heat removal. Similar to using boiling heat transfer at subcritical pressure, where water becomes steam and high heat removal performance is obtained, high heat removal performance can be obtained by heating across the pseudocritical temperature at supercritical pressure.

A supercritical pressure light water reactor is a water-cooled reactor that heats water in a nuclear reactor at supercritical pressure, generating supercritical "steam" for power generation. In subcritical pressure boiling, as seen when heating a "kettle", only a portion of the water becomes steam, leaving water that does not become steam, but when heated at supercritical pressure, all the water becomes supercritical "steam". Both boiling water reactors and pressurized water reactors operate at subcritical pressure, separating the generated steam from the water for power generation, and the separated water is recirculated. The system of the supercritical pressure light water reactor is shown in Fig. 16.2. In a supercritical pressure light water reactor, steam-water separation and recirculation are not necessary. That is, at supercritical pressure, a plant system that sends all the reactor cooling water, called a once-through-flow type, to the turbine for power generation can be realized. Thermal power generation has been practical since the late 1950s with a supercritical pressure, once-through-flow type plant system.

In a supercritical pressure light water reactor, it is not necessary to have a steam-water separator, steam dryer, recirculation system of boiling water reactor, and steam generator of pressurized water reactor. As already mentioned, at supercritical pressure, the enthalpy per unit volume of "steam" is larger than that of subcritical pressure steam, so steam turbines and main steam pipes, etc. become more compact than in light water reactors. Simplification and compactness make the reactor and containment vessel, and the reactor and turbine building smaller than in light water reactors, reducing capital costs. This is the first advantage. The thermal efficiency also improves by about 20% relative to light water reactors, further reducing power generation costs, and reducing the amount of spent fuel and radioactive waste generated per unit of power generated.

Boiling water reactors (BWRs) cool the reactor core using saturated boiling. The mixture of water and steam generated in the reactor core is separated into water and steam by a steam-water separator built into the nuclear reactor pressure vessel, and the steam, from which moisture has been removed by a steam dryer, is sent to a steam turbine to generate electricity. The water separated from the steam is mixed with the water returning from the turbine condenser and is pumped back into the reactor core as cooling water by the recirculation pump. In boiling water light water reactors, only a fraction of the reactor core cooling water becomes steam, with the remainder circulating at high temperatures as recirculated water. Pressurized water reactors (PWRs) cool the fuel rods of the nuclear reactor with nuclear boiling (unsaturated boiling) and use the resulting high-temperature, high-pressure water (332 C) to boil water in a steam generator, generating steam, separating it from the water, and driving a steam turbine to generate electricity. The steam generator has a built-in steam-water separator, and the water separated from the steam is recirculated within the steam generator. Even in pressurized water reactors, the steam used for power generation is saturated steam.

Both boiling water reactors and pressurized water reactors need to recirculate a large amount of water at high temperatures. Since the recirculated water is large in volume and high in temperature, it is not possible to purify the entire amount. From the perspective of solubility, impurities cannot be sufficiently removed unless the temperature is lowered. This is a difficult point in managing the purity of the cooling water of light water reactors, making it difficult to deal with the stress corrosion problems of structural materials such as the reactor core shroud of boiling water reactors and the steam generator tubes of pressurized water reactors. In supercritical pressure light water reactors, like supercritical pressure thermal power plants, after generating electricity with a steam turbine, the reactor core cooling water, which has been cooled down in the condenser, can be fully purified and sent to the nuclear reactor pressure vessel as nuclear reactor cooling water. Supercritical pressure light water reactors have a means of treating nuclear reactor cooling water and can deal with the stress corrosion cracking problems of structural materials that plague light water reactors.

The advantage of supercritical pressure light water reactors is that they can utilize the abundant experience in the design, manufacture, and operation of supercritical pressure thermal power generation equipment. Light water reactors were developed based on subcritical pressure thermal power generation technology up to the 1950s. Thermal power generation then transitioned to supercritical pressure around 1960. Light water reactors, as seen in the Yankee Rowe and Connecticut Yankee in the United States, for example, have been operated at high availability rates from the early power plants. This is thought to be because the technology of power generation equipment such as pumps, valves, piping, and steam turbines is common with thermal power generation, and there was abundant experience in design, manufacture, and operation. The fact that light water reactors are based on the same technical foundation as thermal power generation can be understood, for example, from the fact that there is an association called the Thermal Nuclear Power Technology Association in Japan, and the Japan Society of Mechanical Engineers has a Power Energy System Division, which deals with nuclear power generation technology in the same framework as thermal power generation technology.

Thermal power generation transitioned from subcritical pressure to supercritical pressure in the United States from the late 1950s, and then in Japan and Europe in the 1960s. The world's first commercial supercritical pressure thermal power plant was the Philo Unit 6 in the United States, which started operation in 1957. In Japan, the first supercritical pressure thermal power unit is Tokyo Electric Power Company's Anesaki Unit 1. It started operation in 1967. Since then, ultra-supercritical pressure thermal power, which has higher "steam" temperature and pressure, has been developed and is in operation. In Japan, because coal and oil were imported and fuel costs were high, supercritical pressure thermal power was actively introduced, and high-efficiency ultra-supercritical pressure thermal power plants were developed and constructed. In supercritical pressure thermal power generation, fuel is burned in a structure called a furnace, and boiler tubes are lined up on the wall to heat the water inside. The name of the boiler varies depending on how the tubes are arranged.

"When I started studying nuclear power generation (in the 1950s), I didn't think nuclear power generation would spread rapidly. The reason why light water reactors have become widely popular and have achieved high reliability and availability is because they were able to utilize the equipment and operating experience of thermal power generation", the author heard from one of the pioneers of nuclear power generation in Japan, who was also involved in the design and construction of JPDR, and was a professor at the University of Tokyo. The transition from light water reactors to supercritical pressure light water reactors seems to be a reasonable development of technology.

The author developed the concept of supercritical pressure light water reactors while analyzing and evaluating their characteristics using calculation codes in a research lab at the University of Tokyo since the late 1980s. After retiring from the University of Tokyo, the author conducted research in a research lab at Waseda University from 2010 to 2014. The experiments were conducted with the cooperation of universities and research and development institutions, by applying for the national nuclear system research and development project and obtaining funds. The research results are included in many peer-reviewed papers, and two English books (Oka 2010, Oka 2014) and in the proceedings of the Supercritical Pressure Reactor International Symposium (SCR Symposium), which has been held in rotation in various countries since its first meeting at the University of Tokyo in 2000, among other international conferences and lectures at international institutions (Oka 2011a), and on the homepage of the university research laboratories (Oka 2011b). In 2023, a review paper summarizing research on supercritical pressure light water reactors in Japan was published (Oka 2023). Detailed technical content is described in these documents, so here, the author will introduce the aim of the research, the background, and the history of research and development so far.

The first English book (Oka 2010) describes the reactor core, plant system, plant dynamic characteristics and control, plant startup and stability, safety equipment and safety analysis, and analysis methods and design for fast reactor cores. The concepts and analysis methods are almost the same as those for light water reactor design analysis, except for considering that supercritical pressure light water is a single-phase flow. There are books that describe the design and analysis methods of light water reactors from the perspective of individual fields such as reactor physics and heat transfer fluid dynamics, but there are hardly any books that describe all of these. This English book was created with the intention of also being used as a reference book for the design and analysis methods of light water reactor cores and plants. The analysis methods are based on the methods used in the design and safety analysis of light water reactors, so there is nothing particularly new in terms of methodology. We have developed and used a transient subchannel analysis method for evaluating the maximum temperature of fuel cladding, so the method is not old.

Existing calculation codes can be used for the design and safety analysis of light water reactors, but for the examination of new reactor concepts such as supercritical pressure light water reactors, it is necessary to create your own calculation codes. In that case, the cooperation of experts in reactor physics and heat transfer fluid dynamics is necessary. The analysis of the static characteristics of the core can be done by experts in reactor physics alone, but for the dynamic characteristics of the plant and safety analysis, knowledge of plant control is necessary in addition to the two fields. To go beyond the range of reactor kinetics without feedback learned in reactor physics, it is necessary to create a calculation code based on knowledge of plant control, plant dynamics, and heat transfer fluid dynamics. If a safety system model is added to the plant dynamic characteristic code, plant safety analysis can be performed. The second English book (Oka 2014b) summarizes the results of subsequent conceptual design studies and experimental results on heat transfer fluid dynamics, materials, and water chemistry.

References (Oka 2011a and Oka 2011b) are lecture materials (PowerPoint) from a seminar on supercritical pressure light water reactors held by the IAEA in Trieste, Italy, in 2011. Since it's PowerPoint, it's suitable for understanding the overall picture without spending a lot of time. Furthermore, these lecture materials can also be used as a material to understand the overall picture of the conceptual design and safety analysis method of light water-cooled reactors. While light water reactors require the prevention of boiling transition for the assurance of core integrity, supercritical pressure light water reactors do not experience boiling transition, so the fuel cladding tube temperature becomes the requirement for ensuring core integrity, which is the difference. The design and safety analysis method of the supercritical pressure light water reactor is the same as the analysis method of the light water reactor. As far as the author knows, there is no book that describes the design and safety analysis method of light water reactors in an overview, so these lecture materials should be helpful in understanding them. The formulas used in the analysis are not described in the first lecture material (Oka 2011a), but they are described in the corresponding chapters and sections of the English books (Oka 2010 and Oka 2014b). The reference (Oka 2011b), which explains plant dynamics and control, includes the formulas used in the analysis. The reference collection (Oka 2019) shows the main presentation literature of the "supercritical pressure light water reactor". As already mentioned, 25 years of research on supercritical pressure light water reactors in Japan from 1989 was summarized in 2023 (Oka 2023).

The supercritical pressure light water reactor was selected as the Generation IV nuclear reactor in 2002 and is being researched in Canada, Europe, China, Russia, etc. The trends of each country are introduced on the HP of the secretariat of the Generation IV International Forum (GIF) of OECD/NEA.

Supercritical pressure "steam" is different from superheated steam obtained by heating subcritical pressure steam above its boiling point. Nuclear reactors using superheated steam, as already introduced, have been researched and developed by companies such as GE in the United States in the past, and in the former Soviet Union, they have been operating from the 1960s to the 1980s as Unit 1 (electric output 100 MW, pressure 8.5 MPa, steam temperature 500 C) and Unit 2 (electric output 200 W, pressure 7.3 MPa, steam temperature 501 C) of the Beloyarsk Nuclear Power Plant. However, nuclear reactors using supercritical pressure "steam" for power generation were not well researched until the 1990s when the author started the study.

Research and development of supercritical pressure light water reactors became popular worldwide in the 2000s. In Europe, supercritical pressure light water reactors are being conducted with research and development budgets of the European Community under the name of High Performance Light Water Reactor (HPLWR). In Canada, research on pressure tube type supercritical pressure reactors is being conducted. The status of design research in each country until around 2009 has been published from IAEA (Schulenberg 2011). In China, the Nuclear Power Institute China (NPIC) in Chengdu announced the conceptual design results in 2015 (IAEA 2015). Research is also being conducted at universities (Wu 2022). In Russia, a fast reactor type supercritical pressure light water-cooled reactor is being considered (Baranaev 2004; Glebov 2014). Russia has its own supercritical pressure boiler technology, and research on the heat transfer and fluid flow of supercritical pressure water has been conducted for many years. The PWR type (indirect cycle) supercritical pressure light water reactor was presented by a researcher from the Kurchatov Institute at the ANP'92 international conference held in Tokyo in 1992 under the auspices of the Atomic Energy Society of Japan. This design circulates reactor coolant below the pseudocritical temperature to ensure core moderation. Because it is an indirect cycle SCWR, the coolant temperature at steam generators are high. It poses challenges in terms of materials and corrosion, and the advantages of a simple once-through-flow cycle are lost.

The author has not been involved in the research and development or promotion of supercritical pressure light water reactors since his appointment as Chairman of the Atomic Energy Commission of Japanese government in April 2014. Considering the situation of nuclear power generation in Japan after the TEPCO accident and reflecting on past nuclear research and development in Japan, the author did not promote the development of power reactors, including SCWR, during his tenure as Chairman of the Atomic Energy Commission. The author also thought that the neutrality would be lost if the Chairman of the Atomic Energy Commission promoted a specific type of reactor. Since his appointment as the Chairman, he has shifted his interest to the issues of the interface between nuclear power generation and society.

The concept of the supercritical pressure light water reactor that the author's research group at the University of Tokyo and the group at Waseda University, where he worked afterward, have been researching is called "Super light water reactor" and "Super fast reactor". I will introduce these. The purpose of the research is to improve the competitiveness against thermal power generation, by innovating light water-cooled reactor technology using the technology of supercritical pressure thermal power generation and the experience of light water reactors. The purpose and significance of conducting research at a university is to convey the overall picture of the design and analysis method of light water reactors to graduate students and young researchers through conceptual design research. They need to understand and study by themselves the design and analysis method of the light water reactor, which is an excellent method for human resource development. Although the author studied the basics of structural mechanics and drafting at university, and there are general-purpose finite element method calculation codes that can be used, the author does not have the experience or ability to understand and master the calculation results, design structures, and create manufacturing drawings, so the diagrams drawn in research papers are conceptual diagrams.

The conceptual design study was conducted with the aim of achieving simplicity in design, referencing the experiences of supercritical pressure thermal power and light water reactors. Calculation codes capable of handling supercritical pressure light water cooling conditions were developed and analyzed to determine whether the goal was met. If not met, improvement measures are modeled in the calculation code and analyzed. Judgment criteria for determining results, such as safety, are considered by referencing light water reactors and fast reactors, and it is determined whether they are met. Simpler designs are better, so we start with simple designs and if they are not met, we add equipment and so on. The author himself was able to deepen his understanding of the design and analysis of light water reactors through his research. The research was not possible by the author alone. It was initiated with the cooperation of an excellent associate professor specializing in thermal fluid dynamics and numerical fluid dynamics, and then graduate students, assistants, government-sponsored international students in doctoral programs from abroad, and foreign postdocs hired with research project funds, each of them took charge of their themes and developed calculation codes and conducted research. Because they had the ability to create calculation codes themselves, the research was possible. In addition, as appropriate, the author asked junior colleagues from the Department of Nuclear Engineering at the University of Tokyo, who are researchers at nuclear reactor manufacturers, to listen to the results and give their comments. In particular, their comments on the stability analysis of boiling water light water reactors were helpful. It turns out that water moderation rods affect stability.

The content described below is very specialized, but the author hopes that the readers will understand the history and considerations of research and development, the conceptual design and analysis methods of light water-cooled power reactors, the gap between research and development and practical application, and the difficulties in developing and commercializing new reactors due to the relationship with investment risk. Ideally, a large number of figures should be used for explanation, but they are not included for balance with other chapters and sections. It would be appreciated if you could read it, referring to the figures and explanations in the PowerPoint of the reference (Oka 2011a). There are many common parts between the super light water reactor and the super fast reactor, so in that case, the term "supercritical pressure light water reactor" was used, which means the supercritical pressure light water reactor that the author's research groups at the University of Tokyo and Waseda University studied. The author has been asked which is better, the super light water reactor or the super fast reactor. Although the plant system is the same configuration, the super light water reactor is more realistic because of the uranium fuel.

The pressure of the reactor is 25 MPa, the inlet coolant temperature is 280 C, the outlet coolant average temperature is 500 C, and the core flow rate is about one-tenth of that of a light water reactor. The super light water reactor is a thermal neutron reactor with uranium oxide fuel, and the super fast reactor is a fast reactor with MOX fuel. The supercritical pressure light water reactor is a once-through-flow type reactor, and all the core cooling water is sent to the turbine for power generation. The supercritical pressure light water reactor has a powerful coolant pump, so it can overcome the high pressure loss of the densely arranged fuel rods in the fast reactor core, making it well suited for the fast reactor core. However, to obtain the MOX fuel used in fast reactors, reprocessing and MOX fuel fabrication are necessary. The super light water reactor and the super fast reactor have the same cooling system configuration outside the core, so the results of experiments and tests can be used for both. The super fast reactor has a higher output (power) density than the super light water reactor, and the cladding temperature during accidents is higher, so the safety analysis results of the super fast reactor can also be used as a reference for the super light water reactor.

In the conceptual design study of the "supercritical pressure light water reactor", the core design, plant dynamic characteristics and plant control, plant startup and stability, abnormal transients, and accident analysis were conducted using the calculation codes. Unlike the 1950s when light water reactors were developed, it is now possible to advance conceptual design while quantitatively evaluating the characteristics of nuclear reactors using numerical analysis with computers. The calculation code was used and improved even after graduate students and researchers graduated. There were times when bugs in the calculation code were found when it was handed over, but there were no errors that affected the conclusions of the paper. Core meltdown accident analysis was once conducted using the MELCOR code developed in the United States, but there are no noteworthy results specific to the "supercritical pressure light water reactor", so they will not be introduced here.

In the core design, the reactor pressure was set to 25 MPa and the core inlet temperature to 280 C, and a core that satisfies an average outlet temperature of the core cooling water of 500 C was considered, with the highest fuel cladding temperature during steady operation being around 650 C as a guideline. In light water reactors, the margin for the limit heat flux to prevent boiling transition becomes a guideline for the maximum linear power rating of a fuel rod during steady operation, but in supercritical pressure reactors, there is no boiling transition, so the highest cladding temperature is used as an indicator. These design goals are the same for both the super light water reactor and the super fast reactor. First, the core design and safety analysis results of the super light water reactor will be described, followed by the core design and safety analysis results of the super fast reactor.

The supercritical water reactor is a thermal neutron reactor that uses uranium oxide fuel. Because zirconium alloys lose strength at high temperatures, the fuel cladding is made of stainless steel. The uranium enrichment level is higher than 5% because of the neutron absorption of stainless steel cladding. Zirconium generates a lot of heat when oxidized, which speeds up the rate of core meltdown. The supercritical water reactor, which uses stainless steel cladding, may have advantages over light water reactors with zirconium cladding. The advantages of stainless steel cladding were considered by EPRI after the TMI accident, but during a core meltdown accident, emergency (disaster prevention) response is more important than differences in severe accident behavior due to differences in cladding materials.

Early American light water reactors, such as Yankee Rowe and Connecticut Yankee, used stainless steel cladding until they ceased operation. Therefore, safety standards for abnormal transients and accidents already exist, and these can be used as safety standards for "supercritical pressure light water reactors".

The supercritical water reactor has a low density and high temperature of the core outlet cooling water, so how to ensure the moderation of neutrons is an issue. There are methods either using water rods for moderation or using zirconium hydride, a solid moderator. Our first paper used zirconium hydride as a moderator, but later used water rods. Light water reactors are subcritical pressure, so preventing boiling transition is a constraint on core thermal design, but there is no boiling transition in supercritical pressure light water reactors during rated operation, so the maximum fuel cladding temperature was set as a constraint. It is the same as other reactors of single-phase flow cooling such as sodium-cooled fast reactors.

The core design was carried out by creating a three-dimensional nuclear-thermal coupled core calculation code. Considering the shuffling (replacement) of fuel assemblies for each operating cycle, and also considering the pattern and withdrawal amount of control rods, the core was designed. The maximum linear heat rate is within the limit of 39 kW/m, the maximum cladding temperature is within the limit of 650 °C for each burnup step, there is more than 1% shutdown margin, and it meets the inherent safety (negative output reactivity characteristic). These are used as thermal and nuclear design criteria. The maximum linear heat rate is slightly lower than that of 44 kW/m (BWR) for light water reactors because the average cooling water temperature is high. What is being done in this core calculation is basically the same as what is being done in commercial light water reactors. At that time, a simple method called the node method was used for core management of commercial boiling water light water reactors, but due to the expansion of calculation capabilities and the efforts of excellent doctoral students, it became possible to perform three-dimensional nuclear-thermal hydraulic coupled core calculations by combining diffusion-burnup calculations and thermal hydraulic calculations by single channel models of a fuel rod and a water rod (Oka 2011a, pp 23–31). Before this, calculations were made by modeling the core as a two-dimensional cylinder, so it was not possible to consider changes in the shuffling of fuel assemblies or control rod patterns.

Three types of cores are considered in the calculation, two-pass core, double-tube water rod core, and one-pass core. They are shown in Fig. 16.4. Please refer to the figure and its description first. In the design calculation of the two-pass core, high-density (280C) cooling water flowing in through the upper dome of the nuclear reactor vessel cools the fuel rods and water moderator rods of the fuel assemblies on the outer periphery of the core and the water moderator rods of the fuel assemblies in the center of the core in a downward flow. Then, at the bottom of the nuclear reactor vessel, it mixes with the inlet cooling water flowing into the lower dome of the nuclear reactor vessel along the vessel wall of the nuclear reactor vessel, and then cools the fuel rods of the fuel assemblies in the center of the core. This is called a two-pass core (Oka 2011a p 24, left figure of Fig. 16.4 of this book). With this, it is possible to ensure an average outlet cooling water temperature of 500C, and safety was confirmed as will be described later.

Then, in order to simplify the flow structure inside the reactor, the water moderator rods were made into double tubes, and after the cooling water in the lower plenum of the nuclear reactor vessel flowed through all the water moderator rods, it mixed at the bottom of the core and cooled the fuel rods of the fuel assemblies in an upward flow. A double-tube water moderator rod core was also examined (Oka 2014b p 6, central figure of Fig.16.4 of this book). The siphon break of the double tube can be avoided by making a small hole at the top of the double tube to let out gas.

Finally, one-pass core or single-pass core was analyzed. It adopts single-tube water moderator rods. By surrounding the water moderator rod wall with thin insulation, even if the flow rate of the water moderator rod is small, the water density of the moderator rod can be kept high. Even after the low-density, high-temperature cooling water that cooled the fuel rods and the water that is lower in temperature than the pseudocritical temperature of the water moderator rod are mixed in the outer plenum, it was found that the average cooling water outlet temperature can be kept at 500 C. It is called one-pass core because the cooling water flows only once from the bottom to the top of the core (right figure of Fig. 16.4 of this book). The design results of the one-pass core using a single-tube upward flow water moderator rod are described in the reference (Oka 2014b, Sect. 2.1.1.3). Changes in the maximum linear power and peaking factor for each operating cycle have also been analyzed. The core design and analysis results for the two-pass core are described in the reference (Oka 2011a, slides 17 to 42). The cores of the University of Tokyo, both two-pass and one-pass, have water moderator rods with insulation. Even with insulation, a little heat is transmitted, so the density of the moderating water decreases slightly, and the temperature rise of the fuel rods is mitigated.

The three-dimensional nuclear and thermal-hydraulic coupling calculation provides the maximum cladding temperature for each fuel assembly, but does not indicate the maximum temperature of the fuel rods within the assembly. To find the latter, it is necessary to reconstruct the output of the nuclear fuel rods in the fuel assembly using the results of the three-dimensional core thermal-hydraulic coupling calculation, and calculate the thermal flow within the fuel assembly. This calculation is called a subchannel calculation. By combining the three-dimensional core thermal-hydraulic coupling calculation and the subchannel calculation, the maximum cladding temperature of the fuel rods in all the fuel burnup history and control rod pattern changes was found. As a result, the maximum temperature of the fuel cladding surface was 58 C higher than the maximum cladding temperature averaged for all the fuel in the fuel assembly obtained by the three-dimensional core calculation.

In the calculation to find the maximum cladding temperature, there is uncertainty in the heat transfer coefficient used, the diameter of the fuel rod, the width of the flow path, etc. The method to consider the impact of these uncertainties on the maximum cladding temperature is called the statistical thermal design method, which is used in light water reactor design. When this evaluation is performed, it was found that the impact of uncertainty is 32 C (see pp 39–41 of Reference Oka 2011a).

In other words, when the average outlet temperature of the reactor coolant is 500 C, the maximum cladding surface temperature homogenized for each fuel assembly is 650 C, and the maximum cladding surface temperature in the fuel rods in the fuel assembly is 650 + 58 = 708 C from the results of the subchannel analysis. Adding 32 C due to statistical uncertainty, the maximum cladding surface temperature at rated output is 740 C. The plant safety analysis code calculates the increase in cladding temperature during abnormal transients and accidents, and adds it to 740 C to determine whether it is lower than the cladding temperature criteria during abnormal transients and accidents (see p 42 of Reference Oka 2011a). The subchannel analysis and statistical thermal design method are described in Sects. 2.5 and 2.6 of the reference (Oka 2010), respectively.

Next, the control system and startup system of the plant were designed, and a plant dynamic characteristic calculation code was created and analyzed. Based on this code, the stability during startup and rated operation was evaluated. The analysis method is written in the reference (Oka 2011b). The safety analysis calculation code is a plant dynamic characteristic calculation code with added safety system models. The plant dynamic characteristic calculation model consists of a thermal fluid calculation model that represents fuel rods and water rods in a single channel, a node junction model that represents the conservation of mass and energy of cooling water flowing through equipment and piping in the reactor system by the connection of parts (nodes) of equipment and piping, and a point reactor kinetics model that includes reactivity feedback. The startup of the plant requires analysis over a longer time than the time it takes for the cooling water to return to the reactor system by circulating through the turbine and condenser, so a reactor external circulation model representing the turbine and condenser system is added to these. This system includes a turbine control valve, turbine bypass valve, steam turbine and condenser, condensate desalinator, low-pressure feedwater heater, deaerator, main feedwater pump, and high-pressure feedwater heater. Models representing the valves, pumps, and pressure losses in the piping in this system are used.

The design of the plant control system was carried out using the created plant dynamic characteristic calculation code. First, the step response to the control rod, feedwater (reactor coolant inlet) flow rate, and steam turbine valve opening in the absence of a control system was calculated, and from its sensitivity, it was decided which of the control rod, feedwater flow rate, and valve opening to control for pressure control, main "steam" temperature control, and reactor output, respectively. Then, with the control system attached, the parameters of the PID (proportional-integral-differential) control system were varied in the calculation code to determine the parameters that could be smoothly controlled, and finally, the plant response was confirmed. Details are described in Chapter 4 of the reference (Oka 2010).

Boiling water light water reactors follow the turbine to the reactor power. Pressurized water light water reactors follow the reactor power to the turbine. Thermal power generation is a coordinated control of the turbine and boiler. The control method of the super light water reactor and the super fast reactor is the same as that of the boiling water light water reactor, and the turbine follows the reactor.

The startup method of the plant is either a constant pressure startup method that increases the output of the reactor after supercritical pressure, or a variable pressure startup method that increases the reactor output at subcritical pressure. The analysis method is described in the reference (Oka 2011b from p 38 onwards). It is stated in both cases that the restrictions at startup are that the maximum cladding tube temperature does not exceed the maximum temperature at steady output, and that the density of the steam is low enough not to damage the turbine. In boiling water light water reactors, a temperature rise rate of 55 C/h is set to limit the thermal stress on the reactor pressure vessel, but this limit can be met if time is taken at startup. In the case of constant pressure startup, until the "steam" density becomes low, a pressure reducing valve and a flash tank are provided to bypass the steam so that it does not go to the turbine. In the case of variable pressure startup, a steam-water separator is provided in the main steam line at low output to bypass the steam. In the case of variable pressure startup, boiling transition occurs at subcritical pressure, but if the ratio of flow rate to output is chosen appropriately, the maximum cladding tube temperature can be kept below the maximum temperature limit during steady operation. When the dimensions and weight of the steam-water separator required at startup were determined, the case where variable pressure startup was performed and the steam-water separator was provided in the bypass line had the smallest amount of material.

Thermohydraulic instability analysis and nuclear-thermal coupling instability analysis during startup and rated output operation were performed using a calculation code. The analysis method is described on page 52 and beyond in the reference (Oka 2011b). Equations for the fluctuations were made from the point reactor dynamics, fuel rod heat transfer, water moderator rod heat transfer, fuel channel thermohydraulic, water moderator rod thermohydraulic, and external circulation loop equations. The roots of the characteristic equations were examined by Laplace transformation, and the stability was evaluated from the damping ratio. This method is used in the stability analysis of light water reactors and is called linear stability analysis method. The heat transfer coefficient used included those for supercritical pressure and those corresponding to each boiling mode at subcritical pressure, including film boiling. The critical heat flux was calculated using the table. The criteria for stability judgment are that the damping ratio of thermohydraulic stability (the ratio of oscillation amplitude decay) is 0.5 or less during rated operation and 1.0 or less during startup, and the damping ratio of nuclear thermohydraulic stability is 0.25 or less during rated operation and 1.0 or less during startup. The method and results of startup and stability analysis are described in Chapter 5 of the reference (Oka 2010). The stability analysis was conducted by a doctoral student while receiving comments from a junior colleague at the university who is a researcher at a boiling water light water reactor manufacturer. The comments on the stability analysis including the water moderator rod were particularly useful.

The most important safety indicator for a light water reactor is the water level of the reactor. Ensuring the water level, or the inventory of cooling water, is the basic principle of safety assurance for light water reactors. In boiling water reactors, the water level inside the reactor pressure vessel is monitored, and if it drops, the safety system is activated. Pressurized water reactors have a water level inside the pressurizer, which is used as a signal for the safety system. However, in a "supercritical pressure light water reactor" operating at rated power, there is no water level because the reactor pressure is supercritical. In the "supercritical pressure light water reactor", monitoring and ensuring the flow of cooling water was made the basic principle of safety assurance. There are some water level gauges used in light water reactors that are difficult to accurately measure the water level when depressurization boiling occurs due to pipe rupture accidents, etc. In the TMI accident, due to a misreading of the water-level gauge, the operators understood that the reactor was full of water, stopped the emergency core cooling pump that had been activated to prevent overpressure damage to the primary cooling system, and resulted in core meltdown. Since the primary system of a pressurized water reactor is a closed loop, if it is full of water and the pump does not stop, the system will suffer overpressure damage, so it is stipulated in the operating procedure that the operator must prevent this. It is said that the operators of TMI2 had just received this training. It is not easy to accurately measure the water level when the reactor is depressurized and boiling. Measuring flow is easier and more reliable than measuring water level. In the "supercritical pressure light water reactor", safety system is activated by detecting abnormalities in flow and reactor pressure.

Figure 16.3 shows the plant system and safety system of the "supercritical pressure light water reactor". The Emergency Core Cooling System (ECCS) of the "supercritical pressure light water reactor" consists of a High Pressure Auxiliary Feedwater System (AFS), a Low Pressure Core Injection System (LPCI), and a Safety Relief Valve (SRV)/Automatic Depressurization System (ADS). As a backup for the control rods, there is a Boric Acid Injection System (SLC, standby liquid control system). The main steam line of the containment vessel is equipped with a main steam isolation valve, and the turbine side is equipped with a turbine control valve and a turbine bypass valve (see p 52 of the reference Oka 2011a). These configurations are similar

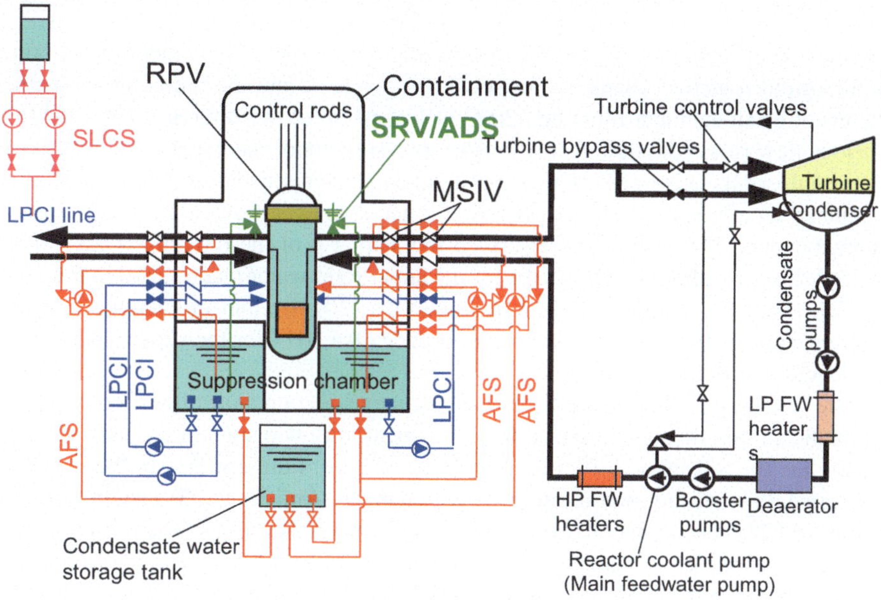

Fig. 16.3 SCWR plant and safety system. *Source* Oka 2011a

to the safety systems of light water reactors. When the cooling water flow rate drops to 90%, a reactor scram (shutdown) signal is given, when it drops to 20%, the High Pressure Auxiliary Feedwater System (AFS) is activated, and when it drops to 6%, the Automatic Depressurization System is opened to depressurize and cool the reactor with the Low Pressure Core Injection System. A high reactor pressure signal stops the reactor at Level 1 and activates the safety relief valve at Level 2. In the case of a low reactor pressure signal, the reactor is stopped at Level 1 and the Automatic Depressurization System (ADS) is activated at Level 2 to cool the reactor with the Low Pressure Core Injection System (see p 53 of the reference Oka 2011a). The capacity of the low-pressure water injection system was determined based on the time it takes to fill the reactor pressure vessel with water and the results of the heat-up of the reactor core as analyzed in a loss of coolant accident. The emergency core cooling system consists of three systems, with the high-pressure auxiliary feedwater system being a turbine-driven pump and the low-pressure core injection system being an electric pump (p 56 of the reference Oka 2011a).

We analyzed the mechanical and overheating damage to the fuel rods, and from the conditions where significant damage does not occur, we derived the criteria for judging during accidents, and from the conditions where systematic damage does not occur, we derived the criteria for judging abnormal transient changes. The upper limit of the cladding tube's maximum temperature was set at 1260 C in the event of an accident and 850 C in the event of an abnormal transient change. In addition to these, there are criteria for reactivity insertion events, which were determined by referring to the criteria of light water reactors. The maximum temperature of the fuel cladding tube at rated output, evaluated by subchannel analysis and statistical thermal design methods, is 740 C, so the temperature rise of the cladding tube due to abnormal transient changes must be 110 C or less, and the temperature rise in the event of an accident must be 520 C or less (p 58 of the reference Oka 2011a). Various flow rate reduction events, reactor pressure abnormal events, and reactivity abnormal events were analyzed using a safety analysis code, and the safety was evaluated by comparing with the criteria for accident events and abnormal transient change events. The results are described in Chapter 6 of the reference Oka (2010) and 2.1.2 of the reference Oka 2014 for the super light water reactor, and in the 59th to 67th PowerPoint slides of the reference Oka (2011b), and in Chapter 7 of the reference Oka (2010) and 2.2.3 of the reference Oka 2014 for the super fast reactor. All meet the criteria.

The "supercritical pressure light water reactor" is a once-through-flow type, so when the automatic depressurization valve opens, a flow occurs in the reactor core, which has the advantage of cooling the reactor core. Furthermore, in the event of a scram failure during abnormal transient changes, Anticipated Transient without Scram (ATWS) can converge even without alternative operations. The reason is that even if the valve is closed and the nuclear reactor scram fails, the flow of cooling water stops in the once-through-flow type reactor, so the increase in cooling water density due to the closure of the valve is suppressed by heating, and the output increase is suppressed. This is because the change in water density due to pressure changes is small at supercritical pressures. Furthermore, the water in the upper dome of the reactor pressure vessel and the water rod play the role of an accumulator (coolant reservoir) in the reactor pressure vessel, mitigating the loss of cooling water events. The behavior during accidents, abnormal transient changes, and ATWS is slower than that of light water reactors. These are safety features and advantages of the "supercritical pressure light water reactor". The safety analysis criterion for ATWS is the same as the accident criterion. In the safety evaluation of ATWS, since the scram failure is considered, there is no need to consider further single failures.

While the loss of coolant accident (LOCA) is an important safety accident event in light water reactors, in the "supercritical pressure light water reactor", the loss of core cooling flow event is an important accident event. This is similar to sodium-cooled fast reactors. The "supercritical pressure light water reactor" is a once-through-flow type (with at least two core cooling systems, although only one coolant line is shown in Fig. 16.2), so the loss of feedwater flow event and the loss of core flow event are the same. In our safety analysis, the total loss of flow event is treated as an accident. The loss of the main feedwater flow of the pressurized water reactor is classified as an abnormal transient event. Pump shaft seizure is classified as an accident. (Please refer to NUREG-0800 NRC Standard review plan 15.0 INTRODUCTION—TRANSIENT AND ACCIDENT ANALYSIS pp 15.0–2 and 15.0–3. Please also note that the feedwater line of a PWR is in the secondary side, not the primary coolant line.)

Whether to treat the total flow loss of the supercritical pressure light water reactor as an accident or an abnormal transient in safety evaluation depends on the equipment design and should be discussed with the regulatory side at the time of application for approval. In the 2005 US SCWR research report (INEEL 2005, pp 104–131), it is treated as an abnormal transient. To meet this criterion, it is proposed to provide an isolation cooling system (isolation condenser, IC) in the bypass line, and it is analyzed. In the research and development report of the supercritical pressure light water reactor (SCPR) conducted by a Japanese reactor manufacturer, it is stated that when it is composed of two completely separated feed and return water systems, it can be treated as an accident with a low frequency of occurrence. As an improvement, it is suggested to treat the abnormality of the return water system as an abnormal transient change, treat the abnormality of the feedwater system as an accident, maintain the feedwater until the reactor stops in the former, and consider a high inertia feedwater pump in the latter (Shioiri et al. 2004).

The reactor pressure vessel of SCWR does not require a steam-water separator or a steam dryer, so it is similar to the reactor pressure vessel of a pressurized water reactor, with a similar shape and dimensions. The weight of the reactor pressure vessel of a pressurized water reactor is lighter than the weight of the steam generator or the weight of the reactor pressure vessel of a boiling water reactor. The inlet cooling water temperature of the reactor is about 280 C, which is almost the same as that of a pressurized water reactor. The inner wall of the reactor pressure vessel is cooled by this inlet cooling water, just like a pressurized water reactor, and since it is not exposed to high temperatures, the same material can be used for the reactor pressure vessel. However, a sleeve is needed at the outlet nozzle of the supercritical steam to relieve thermal stress. The structural design and stress analysis of this part were commissioned to the structural design department of a reactor manufacturer using part of the research funds obtained from the national nuclear system development project, and its feasibility was confirmed. The results of the study are included in the report of the results of the nuclear system development project.

The thickness of the reactor pressure vessel wall, if the diameter of the reactor core is the same, is proportional to the reactor pressure (25 MPa), and is about 60% larger than that of a pressurized water reactor (15.5 MPa), but if the output is set to 1000MW, it will not be much thicker compared to the reactor pressure vessel of the largest PWR, the 1720 MW EPR. The reactor pressure vessel of a boiling water reactor is larger in size and heavier than the reactor pressure vessel of a supercritical water reactor because it incorporates a steam-water separator and a steam dryer.

The reactor control rods, like those of a PWR, are inserted from above the reactor core and fall by gravity when power is lost. This is the same as a PWR. In the diagram of the supercritical pressure light water reactor (JSCWR) made by a Japanese boiling water reactor manufacturer, control rods are drawn that are inserted from the bottom of the core, but a pull-out space for control rod replacement is needed at the bottom of the reactor pressure vessel, so the position of the reactor pressure vessel becomes higher. This makes the containment vessel taller and larger. To insert the control rods from below in the upward direction when power is lost, a highly reliable control rod drive mechanism driven by hydraulic pressure is needed. Such a control rod drive mechanism exists and has been used in BWRs. The JSCWR is a design by a BWR manufacturer. This shows that for nuclear reactor manufacturing companies, their own products and their design and operational experience are important. The control rods of our "supercritical pressure light water reactor" are inserted from the top of the reactor pressure vessel.

The steam turbine will be more compact than the steam turbine of a light water reactor that uses saturated steam. After the steam generates electricity in the turbine, it returns to water in the condenser, and all of it is purified and becomes reactor cooling water. The "steam" conditions are not the same as those of a supercritical pressure thermal power plant, so it is necessary to design according to the steam conditions of the supercritical pressure light water reactor.

The containment vessel can be either a pressure suppression type like a BWR or a large dry containment vessel used in pressurized light water reactors. The latter is considered to have a simpler structure, but in safety analysis, we showed one using a pressure suppression type containment vessel. The containment vessel volume becomes smaller than that of a light water reactor due to the simplification of the system (no need for BWR recirculation system or steam-water separator, no need for PWR steam generator and its piping and pump) and the compactness of the vessel and equipment piping due to the high specific enthalpy of supercritical pressure water. The calculation of power generation efficiency was carried out by creating a heat balance diagram. The nuclear reactor pressure vessel, containment vessel, heat balance, etc. are described in Chapter 3 of the reference (Oka 2010).

Regarding the potential for economic improvement, a comparison with light water reactors in terms of material quantity, etc. for super light water reactors is on page 72 of the reference (Oka2011a), and for European high-performance light water reactors (HPLWR), a comparison of direct and indirect costs with light water reactors has been presented under the title "Potential for Economic Improvement" (Bittermann 2003).

The super fast reactor is a supercritical pressure light water-cooled fast reactor using MOX fuel and stainless steel cladding tubes. The plant configuration except the core is the same as that of the super light water reactor. The core is composed of fuel assemblies with densely bundled fuel rods. When using dense fuel in a high conversion light water reactor, it is necessary to consider the high-pressure loss in the core and the associated instability, but the compatibility with the fast reactor core using dense fuel is good for the super fast reactor because the pump is powerful and the core flow rate is a fraction of that of a light water reactor. The increase in pump power due to the dense core is not a problem. Since the super fast reactor does not require a moderator, it has a higher power density than the super light water reactor. The reactor pressure vessel is small, the capital cost per output can be reduced compared to the super light water reactor, but a reprocessing facility is required to obtain MOX fuel.

Experiments on heat transfer fluid dynamics, cladding material, and water chemistry were commissioned to universities and research and development institutions with a budget obtained by applying for the national nuclear system research and development project under the theme of "Research and Development of Super Fast Reactors". Therefore, the number of research papers presented may be higher for the super fast reactor, but the super light water reactor has an international market for uranium fuel, fewer restrictions, a larger global market, and the importance of improving the competitiveness of nuclear power generation in a deregulated environment, so it is considered to have a higher priority. The experimental results can be used for both the super light water reactor and the super fast reactor.

A unique design challenge for light water-cooled fast reactors is to prevent the coolant void reactivity coefficient from becoming positive when the density of the cooling water decreases, such as during a loss of coolant accident. In high conversion boiling water light water reactors, the core is flattened to make neutron leakage dominant, achieving a negative void reactivity coefficient. However, in super fast reactors, using a flat core increases the diameter of the reactor pressure vessel and proportionally increases the thickness of the vessel wall, which is undesirable. By providing a layer of zirconium hydride about 1 cm thick in the blanket assembly, the fast fission increases during a loss of coolant accident, and the fast neutrons generated in the seed fuel section (MOX fuel section) are slowed down in this layer and absorbed by the Uranium-238 of the blanket fuel, making the coolant void reactivity coefficient negative. This method is being devised and used in super fast reactors. This method and its analysis results are described in the reference (Oka 2010, Sect. 7.3). In small reactors with small cores, the effect of neutron leakage from the core is dominant during a loss of coolant, and the void reactivity coefficient becomes negative.

The core of the super fast reactor was also studied for two-pass and one-pass cores. The two-pass core supplies cooling water from the upper dome of the reactor pressure vessel to the blanket fuel assembly, and after mixing with the remaining reactor cooling water in the lower plenum of the reactor pressure vessel, cools the seed fuel assembly. The one-pass core cools both the seed fuel assembly and the blanket fuel assembly in an upward flow. The former is explained in the references (Oka 2010 Sect. 7.5.6, Oka 2014b Sect. 2.2.1.3, Oka 2011a slides 81 to 83). The one-pass core introduces the design results in the reference (Oka 2014b Sect. 2.2.1.4). Both meet the design constraint conditions with a reactor pressure of 25 MPa and an inlet cooling water temperature of 280 degrees, achieving an outlet average cooling water temperature of 500 C.

The configuration of the safety system of the super fast reactor is the same as that of the super light water reactor. Accidents, abnormal transients, and ATWS are analyzed, and the safety analysis results are in the references (Oka 2010 Sect. 7.11 and Oka 2011a slide 85) for the two-pass core. For the one-pass core, it is described in the reference (Oka 2014b Sect. 2.2.3.4).

In a super fast reactor, even if the fuel rod spacing is 1 mm, it is difficult to achieve a breeding ratio significantly greater than 1.0. The neutron spectrum (energy distribution) of the super fast reactor is softer (the proportion of fast neutrons is relatively small) than that of the sodium-cooled fast reactor, due to its light water cooling and the use of a zirconium hydride layer in the blanket fuel assembly. For accurate calculation of the breeding ratio of the super fast reactor, it is necessary to use group constants that increase the number of low-energy groups, not the group constants used in the core analysis of the sodium-cooled fast reactor. This is described in the reference (Sect. 2.7 of Oka 2014b).

If the blanket fuel assembly can withstand changes in thermal stress during operation by changing the method of bundling fuel rods to a briquette type (a method of placing fuel outside the cooling tube, Tube in shell method), the fuel volume ratio will increase, and it is thought that a breeding ratio significantly exceeding 1.0 can be achieved even with light water cooling. However, changes in temperature and temperature distribution during operation may change the stress on the welds between the assembly wall (Shell) and the cooling tube (Tube), and there is a risk that the welds may break. Although the author has not surveyed thoroughly, the author believes that block-shaped fuel assemblies with metal cooling tubes welded to metal shells have not been used in past power reactors.

Experiments on heat transfer flow of supercritical fluid were conducted by researchers at Kyushu University and the Japan Atomic Energy Research Agency. Kyushu University has a laboratory that has been conducting research on heat transfer flow of supercritical fluid over four generations of professors since the development of supercritical pressure fossil-fired power plants in Japan. The research in this laboratory started with the development of Japan's supercritical pressure fossil-fired power plant, but in recent years it has expanded to applications such as air conditioning. At Kyushu University, experiments were conducted using a supercritical water simulant called HCFC22 (chlorodifluoromethane), including heat transfer and subchannel flow experiments using a single tube, a single electrically heated fuel rod, and three and seven electrically heated fuel rods, turbulence mixing effects of grid spacers, critical heat flux measurement near critical pressure, supercritical fluid condensation test, critical flow, friction pressure loss, cross-flow test of flow path between fuel rods, and verification of coefficients used in core design analysis.

Experiments using supercritical water were conducted at the Japan Atomic Energy Research Agency using a single tube, and the heat transfer measurement results obtained with the simulant were verified. Numerical fluid dynamics analysis of the seven fuel assemblies was also conducted. These results are described in the references (Chapter 3 of Oka 2014b and pp 87–89 of Oka2011a). Detailed experimental results are described in the research report of the Nuclear System Research and Development Project.

Supercritical pressure water is a single-phase flow, making it easier to analyze than the heat transfer flow of a two-phase light water reactor. Using computational fluid dynamics, we analyzed heat transfer degradation phenomena and heat transfer rates, and published the results in a paper. In supercritical water cooling, heat transfer performance degrades at high heat flux or low flow rates. The reasons for the occurrence of heat transfer degradation phenomena were not clarified before. There were the single-phase flow theory and the two-phase flow theory. By using CFD analysis, it has been clarified. In the case of high flow rates, it occurs because the viscous boundary layer near the heat transfer wall thickens and the Prandtl number decreases. In the case of low flow rates, it occurs near the boundary between laminar and turbulent flow, due to the effect of buoyancy making the flow velocity distribution flat and reducing turbulent energy. We also used computational fluid dynamics calculations to determine the heat transfer coefficient and heat transfer degradation heat flux. These results are introduced in the reference (Oka 2010 Sect. 1.3.2). There seem to be no special difficulties in using supercritical water as a reactor coolant.

The supercritical pressure light water reactor is a once-through-flow type, so the core flow rate is about one-tenth of that of a light water reactor. Both the Super Light Water Reactor and the Super Fast Reactor ensure the flow rate for cooling with a fuel rod gap of 1 mm. A new type of fuel rod spacer was devised to accommodate the narrow gap, and its performance was confirmed with computational fluid dynamics. In the design calculations, the heat transfer promotion effect of the spacer is not considered.

The cladding tube material is 15Cr20Ni austenitic stainless steel, and researchers at Tohoku University conducted high-temperature corrosion tests up to 700 C. This stainless steel was developed for Japan's next-generation sodium-cooled fast reactor. Not only flat plates, but also those processed into cylindrical tubes for cladding tubes were tested. The research and development of the insulation material for the water moderator rod was conducted in the nuclear materials laboratory at the University of Tokyo. Yttria-stabilized zirconia (YSZ, Yttria-stabilized zirconia) was developed and tested with its density and porosity as parameters. In the design research at the University of Tokyo, the measured thermal conductivity of the insulating material was used. These experimental results are described in the references (Oka 2014b Chapter 4 and Oka 2011a pp 91–95). Note that yttrium has a larger thermal neutron absorption cross-section than zirconium, so it is better to use zirconia as the insulating material than YSZ.

Experiments in the field of material-coolant interaction and water chemistry were conducted in a laboratory specializing in radiation chemistry at the University of Tokyo. Supercritical water easily dissolves substances. We examined the dissolution and exudation behavior of structural materials into supercritical water. The solubility decreases with temperature and oxygen concentration. Experimental data on the migration and adhesion of corrosion products were also obtained. No peculiar points were observed in the corrosion behavior near the pseudocritical point. These results are described in the references (Chapter 5 of Oka 2014b and pp 96 and 97 of Oka 2011a).

Light water reactors are plagued by stress corrosion cracking problems in the reactor internals and steam generator tubes. It is impossible to treat all the recirculating water, which is at high temperature and pressure, and manage and adjust the water quality of the reactor inlet cooling water. Supercritical pressure power plants are purifying all the condensate. In a supercritical pressure light water reactor of once-trough coolant cycle, all the condensate can be treated. That is, there are means to deal with the corrosion problem of structural materials. The corrosion of structural materials in supercritical pressure power generation is oxidative corrosion. The oxidative corrosion of structural materials in supercritical pressure light water reactors can be dealt with by selecting materials and managing the temperature of the cooling water and its oxygen concentration.

In order to commercialize power reactors, there are issues that go beyond mere technological development and research and development at the interface between nuclear power generation and society. The author believes that the experience with the research and development of the supercritical pressure light water reactors will be a reference for future research and development. The contents described below are based on facts, and the author's considerations and opinions based on them are also considered to be useful references, so they are described as a story.

The idea of a supercritical pressure light water reactor came to the author in the late 1980s when supervising graduate students at the University of Tokyo. We were studying the concept of a steam-cooled fast reactor and wanted to somehow make the void coefficient negative in the event of coolant loss. When the author told a graduate student who was doing the calculations to increase the density of the steam, he said there was no pressure. When the author asked why, he said the steam table he was using was for subcritical pressure and there was no data for supercritical pressure on the table. That's what he meant by no pressure. That's when the author learned that there is a state called supercritical pressure. The author immediately thought that if we could achieve supercritical pressure, we could simplify the boiling water type light water reactor. Upon investigation, the author found that supercritical pressure was already being used in thermal power generation. The fact that the author had heard from his university mentor that light water reactors were developed based on thermal power generation technology, and that at the time the author was attending a study group on passive safety light water reactors outside the university and hearing discussions about improving light water reactors, can also be cited as background to this thinking.

Although we had been publishing results in peer-reviewed papers since around 1991, the first comprehensive presentation was made in 1992 at the ANP92 international conference held in Tokyo, sponsored by the Atomic Energy Society of Japan. At this time, we presented the concept of a direct-cycle supercritical pressure light water reactor, while researchers from the Kurchatov Institute in Russia presented the concept of an indirect-cycle supercritical pressure light water reactor. The first time our research attracted international attention was in 1996 at the Pacific Basin Nuclear Conference held in Kobe, where we presented a paper on the potential for reducing the mass and improving the thermal efficiency of supercritical pressure light water reactors. The former chief engineer of Canada's AECL took notice of its advantages. Later, the former vice president of a large power company in Japan also took notice. These two individuals were involved in the early development and construction of CANDU and light water reactors. Since then, Canada has been active as a member country of the Generation IV International Forum up to the present. Unfortunately, both of them had already retired from their leadership positions at the time. When the author met them later, they told me, "I'm no longer active." The author didn't ask for details, but the author later realized that this phrase had various meanings.

In the late 1990s, when the author visited the Karlsruhe Institute in Germany, the author was asked by the head and professor of the nuclear energy department (at that time, there were a few professors in Germany) to cooperate in research on supercritical pressure light water reactors. This professor was nearing retirement and seemed interested in supercritical pressure light water reactors as a future research theme for the institute. After this professor's retirement, under the newly appointed professor (head) of the Karlsruhe Institute of Technology (which was reorganized from a federal research institute to a university), research on supercritical pressure light water reactors was conducted under the name of high-performance light water reactors (HPLWRs), with the budget of the 5th Framework Program (1998–2002) and the 6th Framework Program (2002–2006) of the European Community, with participation from Germany, Finland, and other European countries. The author was invited to and attended the annual summary meeting. A technician from Areva in Germany participated in this meeting, but Areva in Germany did not intend to develop supercritical pressure light water reactors. As a reactor manufacturer, they seemed to have a policy of cooperating with the projects of their own country's nuclear research institutes, not limited to this research. This technician was an expert in severe accidents, and it was good to get to know him through participation in the high-performance light water reactor meeting. Currently, research on supercritical pressure light water reactors in Europe is being continued since September 2020 under the project name ECCsmart of EURATOM (European Atomic Energy Community), with funding from the EC and participation from 20 institutions and universities in 13 European countries, Canada, and China. This project sets technical, strategic, and regulatory goals, with a combination of climate and environmental change response, supercritical water technology, and small modular reactor technology as its elements.

In the late 1990s, the Idaho National Engineering and Environmental Researchers (INEEL, now INL, Idaho National Laboratory since 2005) asked for cooperation in the research of supercritical pressure light water reactors, offering a reward. I replied that I would cooperate without needing a reward. It might have been normal in Japan to cooperate without compensation, but it was not good to reply that no reward was needed. In the United States, you pay for what you ask for. This makes responsibilities clear. In the United States, researchers from INEEL, with the cooperation of material corrosion experts from U.S. universities and technicians from light water reactor manufacturers, obtained competitive funding called Nuclear Energy Research Initiative (NERI) from the U.S. Department of Energy, and conducted research and development on core design, fuel and structural materials, plant engineering, and safety analysis for four years from 2001 (INEEL 2005, 2003). INEEL's research referred to the design of the University of Tokyo, but the author did not participate in the research.

The core was initially considered to use a core with a zirconium hydrides moderator, and then a water-moderated rod core was considered based on the design that the University of Tokyo was calling SCLWR-H at the time. INEEL researchers proposed a power channel core in the second year research report (INEEL 2003). The power channel core is an idea that bundles 18 fuel rods into a hexagonal fuel assembly, puts them in an insulated zirconium alloy channel box (wrapper tube), inserts a Y-shaped control rod in the gap between the assemblies, and the cooling water in this gap also slows down the neutrons. The cooling water descends the gap, mixes at the bottom dome of the reactor pressure vessel, and cools the fuel assembly in an upward flow (INEEL 2003, pp 13–14)). However, what is mainly analyzed in the final report is the water-moderated rod core.

The plant engineering and safety analysis of Idaho Lab's SCWR research is being conducted by Westinghouse (WH) (INEEL 2005, 2003). A subchannel analysis of the fuel assembly (fuel rod diameter 10.2 mm, pitch 11.2 mm, gap width 1 mm) of the SCLWR-H super light water reactor from the early 2000s at the University of Tokyo is being conducted using the VIPRE-W code. The results show a very large increase in the temperature of the fuel cladding tube. This calculation was done under the assumption that the water moderator rod wall is adiabatic because the budget constraints did not allow for modifications to the calculation code. In other words, heat transfer between the water moderator rod and the cooling water is not considered. Furthermore, the results of the three-dimensional nuclear-thermal calculation are not used as input for the subchannel analysis, and the concentration distribution of the fuel rods is not optimized. The hot channel factor is also not considered. It is stated that the impact of factors such as deformation of the fuel rods is significant due to the narrow gap width of the fuel, but statistical thermal design considering the manufacturing tolerance of the fuel rods and the error of the heat transfer coefficient is not conducted due to budget constraints. In conclusion, it is stated that there is a high risk in further considering the SCWR because the narrow gap between the fuel rods has a significant impact on the cladding tube temperature.

The calculation of WH's VIPRE-W is an analysis under extreme conditions where no heat is transferred to the water moderator rod at all. Even if the water moderator rod is insulated, some heat is transferred, so the assumption that a 4 m long water moderator rod does not contribute at all to the cooling of the adjacent fuel rods is extreme and not reasonable. The VIPRE-W calculation code is a calculation code used for subchannel analysis in PWRs. Water moderator rods are not used in PWRs. As we have already mentioned, we use the results of three-dimensional core nuclear-thermal coupling calculations, perform subchannel analysis, and further consider the impact of error uncertainty in statistical thermal design. Our subchannel analysis results do not deny the feasibility of the concept. The water moderator rods of our super light water reactor's two-pass core and one-pass core are insulated. In the reference (Oka 2010), the results of the three-dimensional nuclear coupling core design in Sect. 2.4, the results of the subchannel analysis in Sect. 2.5, and the results of the statistical thermal design in Sect. 2.6 are for insulated water moderator rods (see pp 20–42 in the reference Oka 2011a). Care must be taken regarding the application range of the heat transfer correlation formula for supercritical fluids used in subchannel analysis, etc. Initially, we used the Koshizuka-Oka correlation formula created by numerical fluid dynamics calculations, but since it underestimates the heat transfer at low flow rates, we are now using the Watts correlation formula. The results in the reference (Oka 2011a) are based on the Watts correlation formula. The Koshizuka-Oka correlation formula was created with the intention of using it for safety analysis, so it asymptotically approaches laminar heat transfer at low flow rates. Therefore, since it underestimates heat transfer at low flow rates, it cannot be applied to the calculation of heat transfer of water moderator rods.

The above analysis of WH is poor, and the conclusion is not logically valid. It is illogical to conclude that the feasibility of a supercritical light water reactor is nonexistent, by stating that the cladding temperature becomes too high under extreme conditions where heat does not transfer to the water moderator rod without optimizing the power distribution of the core and fuel assembly. If it is to be deemed inadequate, the power of the core and fuel assembly should be optimized, and the analysis should be conducted under conditions where heat is transferred to the water moderator rod through insulation, and the conclusion should be stated. This is a logical method of examination.

According to Innovation Dynamics, which studies the history of technological innovation, innovation does not come from companies that have products dominating the market (Utterback 1994). The supercritical pressure light water reactor may potentially compete with WH's PWR product. There is a systemic problem in the evaluation method itself by involving such a company in the evaluation of the supercritical pressure light water reactor. Nowadays, knowledge about supercritical pressure light water reactors is much more abundant than in the early 2000s. If the United States has pride in leading the world's use of nuclear power, it should reconsider the supercritical pressure light water reactor and review this INEEL evaluation. In the United States, the economic superiority of light water reactors is being threatened, and the need for them is high.

In Idaho Lab's SCWR research, researchers at Idaho Lab have modified the RELAP5 code so that it can analyze supercritical water cooling, and are conducting abnormal transient and accident analysis. WH proposes that an isolation condenser (IC), which operates in emergencies, should be installed as a countermeasure because the loss of main feedwater flow is an abnormal transient change and the increase in cladding temperature is large. The isolation condenser is a passive device, so it is easy to maintain and manage as a safety system. This proposal is interesting. However, as already mentioned, whether the loss of main feedwater flow (loss of all feedwater flow) is considered an accident event or an abnormal transient event depends on the equipment design and should be decided through discussions with the regulatory side. The fuel cladding of the core examined by Idaho Lab is ODS steel (Oxide Dispersion Strengthened Ferritic Steel) (INEEL, 2005). This is believed to reflect the opinion of American material experts that stress corrosion cracking of austenitic stainless steel becomes more severe at high temperatures. This expert has compiled a review on the materials for supercritical pressure light water reactors and published it in 2007 (Was 2007). "Ferrite martensitic steel shows good resistance to stress corrosion cracking, but is poor in terms of oxidation corrosion. Austenitic stainless steel and nickel-based alloys are strong against oxidation corrosion, but weak against stress corrosion cracking. We also discussed the design of material grain boundaries and surface modification", it says.

The American result that stress corrosion cracking becomes more severe at high temperatures in austenitic stainless steel contradicts the experimental results of Japanese reactor manufacturers that austenitic steel (sensitized 304 steel and 316L steel, solution-treated 316L steel and 310S steel) loses susceptibility to stress corrosion cracking at high temperatures (Was 2007, p 30, Shioiri 2003, Fig. 13). The author is not a material expert, so it is not clear why the conclusions are opposite, but supercritical pressure light water reactors can purify all the cooling water. Light water reactors circulate most of the cooling water at high temperatures through the core, and cannot purify all of it. It is not appropriate to consider stress corrosion cracking from the experience of light water reactors without considering this difference. Corrosion tests should be conducted under once through-flow conditions where all the return water is treated at low temperatures in the test loop.

The supercritical pressure light water reactor was selected as a Generation IV reactor in 2001. However, the United States did not participate in the international research program for SCWR (System Agreements, GIF-SCWR). The United States participated in the sodium-cooled fast reactor (SFR, sodium fast reactor) and the very high temperature reactor (VHTR, very high temperature reactor). The reason for not participating is not clear, but in INEEL's SCWR research, WH expressed a negative opinion based on the subchannel analysis results, and the sodium-cooled fast reactor and high-temperature gas reactor had been researched and developed in the United States in the past, and there were past design results and organizational research promotion groups, and the industry, such as reactor manufacturers, was not interested because the new construction of light water reactors was important. At that time, the Generation III reactors were the reactors to be built, and the Generation IV reactors were the reactors to be researched. Since then, the situation surrounding light water reactors in the United States has changed significantly due to the delay in the construction of Vogle 3 and 4 and the shale gas revolution, which has lowered natural gas prices, raising questions about the competitiveness of light water reactors. The investment environment for nuclear power plants varies from country to country. Currently, the United States is the country that most needs research and development of supercritical pressure light water reactors. Currently, Canada, Euratom, Japan, China, and Russia are participating in the Generation IV SCWR research program. Russia joined in 2011 and China in 2014.

Next it discusses the research and development of supercritical pressure light water reactors in the industrial sector in Japan. Interest in supercritical pressure light water reactors increased globally toward the end of the 1990s. At that time, the heads of a Japanese nuclear reactor manufacturer expressed their desire to develop supercritical pressure light water reactors, to which the author responded affirmatively. The author once suggested aiming to obtain design certification from the U.S. NRC, but they said that would be a daunting task. This company just acquired the U.S.'s WH. The author seems to recall them saying they wanted to consider it as a successor to the AP1000, but not sure. The author doesn't know whether this manufacturer seriously wanted to develop and commercialize it in the future, or just wanted to research it, i.e. secure national research funding and maintain the capabilities of their researchers. In any case, this nuclear reactor manufacturer obtained competitive funding from the ministry in charge of the nuclear industry, and with the cooperation of other nuclear reactor manufacturers and multiple university laboratories, conducted research and development for five years from 2000 on plant concepts, heat transfer, and materials sub-themes, referred to as SCPR (Shioiri et al. 2004).

The plant concept is based on the core design of the super light water reactor called SCLWR-H at the University of Tokyo at the time, and includes fuel integrity, core nuclear, thermal hydraulic, and structural design. Heat transfer experiments are being conducted at Kyushu University, focusing on transient heat transfer characteristics. Material tests are being conducted using improved materials of austenitic stainless steel (316L, 310S), including irradiation characteristics tests under ion and electron beam irradiation, corrosion tests, and stress corrosion cracking susceptibility evaluation tests. Tests and evaluations are also being conducted on Ni-based alloys and Ti alloys. The overall summary was presented in two references (Shioiri 2004, 2003). Subchannel analysis is being conducted for the core design, along with comparisons with calculations by the general-purpose computational fluid dynamics code STAR-CD, and three-dimensional nuclear-thermal design of the core. The company's research continued even after the five-year project, with core design, heat transfer fluid tests, and material tests being conducted as collaborative research for GIF. The research results from this period are only available in international conference papers, so the details are unclear. Regarding the core design, three-dimensional nuclear-thermal design, subchannel analysis, and statistical thermal design have been conducted, and a core using a fuel assembly with a large square water moderator rod surrounded by fuel rods has been presented as JSCWR. The author will introduce this core later.

Research centered on this company was discontinued in 2011 due to budget cuts following the Great East Japan Earthquake, and has not been conducted since. The company suffered a significant loss in the 2000s due to delays in the construction of the nuclear power plant in the United States by WH, which it had acquired. The prospects for the construction of nuclear power plants in the United States also drastically decreased due to the construction of gas turbine power plants, which was a major blow. In any case, it is understood that the burden of development was high for a private Japanese company to commercialize a supercritical pressure light water reactor, compared to market prospects. For example, it is necessary to have one's own heat transfer fluid test equipment and to develop calculation codes for plant design and safety analysis. Furthermore, there are tasks that are a heavy burden for private companies, such as obtaining permits and designing and manufacturing equipment and plants. This company is not a manufacturer of supercritical pressure boilers, so the experience of designing and manufacturing equipment was only available to other companies. Furthermore, in order to commercialize it, it is necessary to have a market prospect of being able to construct about 10 units. These are the challenges when trying to commercialize a supercritical pressure light water reactor. It is difficult to meet these requirements from a perspective confined to the domestic market. It may have been unreasonable to try to develop it only with private companies, without seeking cooperation from national research and development institutions and other ministries, due to the influence of Japan's bureaucratic sectionalism. However, it is a fact that there was a reactor manufacturer who focused on the advantages of supercritical pressure light water reactors and tried to develop them, even though they did not sell them.

The 2000s were a time known as the nuclear renaissance, and a competitive fund called the "Nuclear System Research and Development Project" was established in relation to the Ministry of Education, Culture, Sports, Science and Technology. In 2005, the author applied as the research representative under the title "Research and Development on Light Water Cooled Super Fast Reactor", won the competitive fund, and conducted research on the design of the reactor core and plant concept, heat transfer fluid experiment of the reactor core, cladding material development and water chemistry, etc. with the cooperation of researchers from related laboratories of the University of Tokyo, Japan Atomic Energy Agency, Kyushu University, Tohoku

University, etc. The first phase (2005–2011) was conducted at the University of Tokyo, and the second phase (2012–2015) was conducted at Waseda University. Both the first and second phases were evaluated by the evaluation committee of this competitive fund, and the results have been highly evaluated. The research results have already been introduced. The core design and safety analysis were carried out by postdocs hired with the acquired funds, who were divided into themes and created and analyzed calculation codes. Since Japan has a lifetime employment system, there were hardly any postdocs in the field of nuclear reactor engineering, so we hired postdocs who had graduated from doctoral programs at foreign universities. Many of them needed two years of overseas experience to become university professors in their home countries, and many of them were excellent. They are now active as faculty members in the nuclear-related departments of universities in their home countries. In addition, the research on the super light water reactor (thermal neutron reactor) only involved creating calculation codes and examining reactor concepts, so there was no need to acquire external research funds. This was possible because there were excellent young faculty members and graduate students. If a private company tries to do the same thing with its researchers, it would require high salaries, which is not feasible.

There are two reflections on the research during this period. One is that the Japan Atomic Energy Agency should have developed and maintained the calculation codes necessary for the design analysis of the supercritical pressure light water reactor with the competitive funds we acquired. The other is that we did not create an organizational network for research and development, including the Japan Atomic Energy Agency and nuclear reactor manufacturers. Looking at the SCWR research at INEEL in the United States, INEEL researchers are maintaining and analyzing the RELAP5 calculation code so that it can be used for SCWR safety analysis. This calculation code can now be used worldwide. The calculation codes necessary for the design analysis of the supercritical light water reactor include not only the safety analysis code and the three-dimensional core thermal coupling nuclear-thermal calculation code, but also the subchannel analysis code, statistical thermal design method, plant control and stability analysis code, fuel behavior analysis code, etc. If these calculation codes developed for conceptual design research could have been maintained in Japan as an extension of the publicly available light water reactor design analysis calculation codes, it would have been useful not only for the foundation of future supercritical pressure light water reactor research and development, but also for the development of personnel who understand the design and analysis of light water reactor plants and the construction of a knowledge base.

Regarding the creation of the network, one of the reasons was that the author did not fully understand the state of research and development by the reactor manufacturers who said they wanted to develop it. It's an excuse, but at that time, the author was leading the establishment of the University of Tokyo's Department of Nuclear Engineering and Management and Nuclear Professional School based on the faculty quota of the Nuclear Engineering Research Laboratory of the University of Tokyo, which the author had been in charge of for many years. The author proposed and edited the creation of a series of nuclear engineering textbooks of Ohmsha in Japanese and Advanced Course in Nuclear Engineering of Springer in English based on the educational curriculum of the Nuclear Professional School. As a representative, the author won competitive funding from the Ministry of Education, Culture, Sports, Science and Technology's Global Center of Excellence (COE) program in the field of nuclear power and led its activities. After moving to Waseda University, the author had to spend a lot of time launching and operating its Cooperative Major in Nuclear Engineering. In any case, it is a point of reflection that the author did not have time to consider the future of the research and development of the supercritical pressure light water reactor. It's embarrassing, but the author read the Idaho Research Institute's 2005 report and the Japanese reactor manufacturer's SCPR research paper in 2022, when the author wrote this book in 2022. In addition, the author could not and did not intend to commercialize the supercritical pressure light water reactor because the author has no experience in design and manufacturing.

One of the good things about the supercritical pressure light water reactor being selected as the Generation IV reactor is that our research has become world-famous, and many excellent government-sponsored students (students studying for a doctorate) who have won scholarships in their own countries have come to our laboratory. The postdocs we hired with the competitive funds we won were also advantageous in recruiting because of the high international reputation of the research theme, and many of them were excellent. The problem was that the author accepted the role of Japan's representative for GIF-SCWR. The author was asked by Japanese reactor manufacturers who said they wanted to develop a supercritical pressure light water reactor to take on the role of Japan's representative for GIF-SCWR because it could not be done by a private company, and I accepted because they said they would cooperate, but the GIF meetings were mostly administrative and there was no point in going abroad to participate as a busy university professor. In other countries participating in the SCWR International Forum, government officials and researchers from research and development institutions are in charge and participate. In this respect, Japan's system for GIF-SCWR was extremely fragile and disappointing, relying on individual university professors on top of the vertical division of government agencies, and there was no support. In retrospect, it would have been better to have a representative from a research institution. University professors need to run their laboratories and provide university education almost by themselves, and are extremely busy. Now, you can participate in meetings on the Internet without having to travel abroad.

The concept of the supercritical water-cooled reactor announced by around 1960 was introduced at the first symposium on supercritical water-cooled reactors held at the University of Tokyo in 2000. At that time, the method was to search for literature in the library, but now it is easy to search for literature on the Internet. Based on the newly found report, the author will introduce the study of the supercritical pressure light water reactor in the United States in the 1950s.

The Argonne National Laboratory (ANL) in the United States conducted reviews of the following three studies in 1960 (Marchaterre 1960).

1. Pratt and Whitney Aircraft Study

2. WCAP Design Study
3. Hanford Conceptual Design.

1 is a ducted boiler propulsion for aircraft propulsion. Development of fuel and materials is underway.

2 is a design study of a commercial marine reactor by WH. It is a reactor using a pressure vessel, but an indirect cycle is selected. There are two flows within the reactor vessel. Water at 4000 psia (27.6 MPa) and average temperature of 500 F (260 C) is used for a moderator. Supercritical steam being heated from 860 F (460 C) to 1000 F (538 C) cools the fuel assemblies. Since an indirect cycle was selected, a high-temperature, high-pressure compressor (blower) is required to circulate the low-density supercritical "steam" above the pseudocritical temperature after heat exchange in the steam generator, and this equipment has never been made, according to the ANL report.

3 is a heavy water-moderated light water-cooled pressure tube type reactor, which loads fuel elements into the pressure tube and cools it with supercritical pressure water. The shape of the fuel element is complex. Refining this reactor concept leads to Canada's CANDU-SCWR.

At that time, it was when the first supercritical pressure thermal power plant, the Philo Plant, started operation in the United States. Valves, pipes, turbines, feedwater pumps, and feedwater heaters have experience if it is 5000 psi (34.4 MPa), 1400 F (760 C) or less. The ANL report states that a supercritical pressure light water reactor using this technology is reasonable as the development of a water-cooled power reactor. It also states that the greatest advantage is the improvement in thermal efficiency, but the author cannot agree with this. Since the proportion of fuel costs in the power generation cost is large in thermal power generation, the improvement in thermal efficiency has a great effect on reducing power generation costs, but the proportion of capital costs is large in nuclear power generation. Therefore, the reduction in capital costs due to the compactness of the equipment by using supercritical "steam" with a higher enthalpy per volume than subcritical pressure steam is more important than improving thermal efficiency.

There are several U.S. patents related to supercritical pressure water-cooled reactors in the 1960s, including a patent granted in 1969 by an employee of Siemens in Germany (Winkler 1969).

This reactor is a pressure vessel type, but it differs from the super light water reactor in that there is a region of moderator water surrounding the outside of the cylindrical fuel rod cooling channels. The cooling of the fuel rods is by downward flow, and the water supply for fuel cooling and moderation is connected to the nuclear reactor pressure vessel by separate pipes. The system outside the nuclear pressure vessel is not described. In addition, both this German patent design and the WH design introduced in the 1960 report from Argonne National Laboratory have a main "steam" pipe connected to the bottom of the nuclear reactor pressure vessel. Current light water reactors need to place inlet and outlet pipes above the core from the perspective of securing the water inventory inside the nuclear reactor pressure vessel in the event of a loss of cooling water due to a cooling pipe rupture.

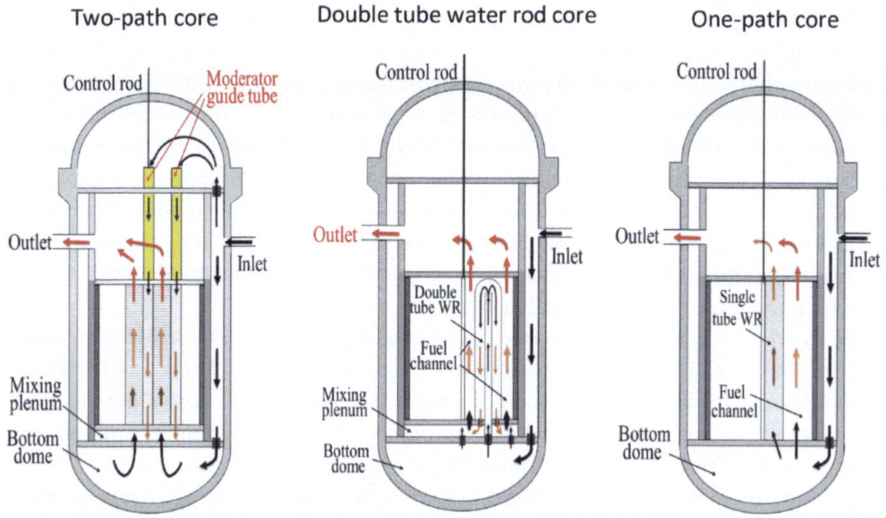

Fig. 16.4 Three types of SCWR cores. *Source* Oka 2014

Various cores have been studied for the supercritical light water reactor. The main ones and their features are summarized. The inlet temperature of the cooling water in the core is 280 C, and the average outlet temperature is 500C, which is the same for all cores of ours, European HPLWR, INEEL, and Chinese NPIC.

- Zirconium hydride-moderated core: It is a solid moderator and does not contribute to the cooling of the fuel rods, so when a subchannel analysis is performed, depending on the fuel assembly design, the temperature rise of the fuel cladding in the high output channel may be large. Since there is no need for a structure to supply water to the water moderator rod, the in-reactor structure is simplified.

  The three types of cores of the super light water reactor are shown in Fig. 16.4.

- Two-path (pass) core of the super light water reactor. The cooling water that enters the nuclear reactor pressure vessel is divided into those that go to the upper dome of the nuclear reactor pressure vessel and those that go to the lower dome of the nuclear reactor pressure vessel through the downcomer. The cooling water in the upper dome descends the water moderator rod, mixes with the cooling water flowing in from the lower dome in the mixing plenum below the reactor core, and cools the fuel rods by upward flow. The water moderator rod is insulated to keep the water density high. The analysis results shown in the reference (Oka 2011a) are for this core. Even if it is insulated, some heat is transmitted to the water moderator rod, so the temperature rise of the fuel cladding is mitigated and the design criteria are met. The structure of the water supply part to the water moderator rod is complex.

- Double-tube (pipe) water moderator rod core of the super light water reactor: The cooling water entered into the reactor pressure vessel descends down the downcomer and flows into the inner pipe of the double-pipe water moderator rod from the lower dome. After rising up the inner pipe, the cooling water turns back at the top, descends down the outer pipe, and mixes with the cooling water coming from the lower dome in the mixing plenum below the reactor core to cool the fuel rods in an upward flow. In this core, it is necessary to have separate cooling water inlet plenums for the water moderator rods and the fuel rods at the bottom of the core. The siphon break of the double pipe can be prevented by drilling a small hole at the top of the pipe to release gas.
- One-path (pass) core of the super light water reactor: Cooling water is supplied from the lower plenum to the water moderator rods and fuel rods, and both are cooled in an upward flow. The cooling water flowing from the top of the insulated single-pipe water moderator rod is released into the outlet plenum and mixed with the cooling water of the fuel channel. By insulating the water moderator rods and reducing their flow rate, it is possible to prevent a decrease in the average outlet temperature after mixing, while ensuring that the outlet temperature of the water moderator rods does not exceed the pseudocritical temperature, as shown in the core design. The in-core flow is most simplified.
- Three-pass core of the European HPLWR (IAEA 2011): 50% of the cooling water descends the water moderator rods of the fuel assemblies through the control rod guide tubes from the upper head. It then turns back, rises up the outside of the fuel assemblies, descends inside the core shielding in the circumferential direction, mixes with the remaining 50% of the water in the lower plenum, and cools each of the radially divided fuel assemblies (52 each) in turn in an upward flow, downward flow, and upward flow.
- INEEL's power channel core (INEEL 2003, pp 13–31): Hexagonal fuel assemblies, with Y-shaped control rods inserted from the bottom of the core into the water region between the assemblies. The cooling water descends here, enters the lower plenum, and rises up inside the fuel assemblies. The hexagonal fuel assemblies are referred to as power channels. The channel box is insulated, and it is stated that zircaloy alloy can be used in terms of temperature. It is a two-pass core.
- Two-pass core of China's NPIC: A thick single water moderator rod is surrounded by two rows of fuel rods in a square fuel assembly, and a cross-shaped control rod is inserted between four fuel assemblies. The cooling water flows down the fuel rods in the central fuel assembly, all the water moderator rods in the fuel assemblies, and the gaps between the fuel assemblies, mixes in the lower plenum, and cools the fuel rods in the peripheral fuel assemblies in an upward flow (IAEA 2015).

- JSCWR Reactor Core of Japanese BWR Manufacturer: A large fuel assembly with a thick, square water moderator rod in the center, surrounded by four rows of fuel rods. It is a two-pass core with a downward flow of the water moderator rod and an upward flow cooling of the fuel rods. There is a water area on the outside of the fuel assembly where cross-shaped control rods are inserted. Cooling water is supplied from this water area to the upper part of the water moderator rod, and the descending cooling water is supplied to the lower part of the fuel assembly to cool the fuel rods in an upward flow. It aims to equalize the power distribution using several types of enriched uranium ranging from 5 to 10% and the burnable poison gadolinia. The flow distribution of the fuel assembly is also of 10 types. Out-in fuel exchange. Cooling water inlet temperature 290 C, outlet temperature 510 C. Statistical thermal design and subchannel analysis are performed, and it is evaluated that there is a temperature rise of 78 C. Including this, the maximum cladding temperature is 700 C. The gap between fuel rods is 1.5 mm, while many other designs are 1 mm. The material of the fuel cladding tube and the channel box is 316 stainless steel (Sakurai 2011). The effective length of the fuel is 4.2 m, and the length of the fuel rod is 5.8m. JSCWR inserts cross-shaped control rods from the bottom of the core (Yamada 2011). In the analysis results of the two-pass core of the University of Tokyo introduced in the reference (Oka 2011a), the maximum cladding temperature considering subchannel analysis and statistical thermal design is 740 C, but in this core, by using a large number of enrichment splits and detailed flow distribution, the power distribution and fuel cladding tube temperature distribution are flattened, and the maximum cladding temperature considering subchannel analysis and statistical thermal design is improved to 700 C. The JSCWR method of inserting control rods from the bottom of the reactor pressure vessel requires space for pulling out the control rod drive shaft at the bottom of the reactor pressure vessel, which raises the position of the reactor pressure vessel, increases the containment vessel height, and requires a hydraulic drive system for control rod operation, which is not good. As already mentioned, it is a core designed from the perspective of using as many BWR manufacturer's equipment as possible.

In any case, the development elements of the supercritical pressure light water reactor lie in its fuel assembly. Since the fuel assembly is replaced after being used in the core for a certain period, it can be improved sequentially like the fuel assembly of a light water reactor. Since the supercritical pressure light water reactor can use the equipment and experience of light water reactors and supercritical pressure thermal power plants other than the core, it is considered that there are no major development elements.

Furthermore, the SCWR in Canada was initially a CANDU type using pressure tubes, but later, a design using a pressure vessel was also proposed. In the design of the pressure vessel type, the moderator is heavy water in a low-pressure vessel, and many pressure tubes are inserted into the lower heavy water vessel from the bottom of the pressure vessel that stores the reactor cooling water above it. The cooling water descends inside the pressure tube, turns back at the bottom, and cools the fuel rods arranged in the annular part (Yetisir 2016).

There are errors and misunderstandings in the explanations and books published in English on the Internet about the supercritical pressure light water reactor. Some have already been deleted, but they are introduced in a Q&A format. W is an explanation of the error, C is a lack of understanding, and A is the answer and the author's opinion.

W: The supercritical pressure light water reactor has been in practical use at the Beloyarsk (Beloyarsk) power plant in Russia for a long time. The Beloyarsk power plant is a power plant that uses supercritical "steam". (This error was left unattended in the description of SCWR on Wikipedia for many years. There were also mistakes in English books.)

A: The Beloyarsk power plant Units 1 and 2 use superheated steam and are not supercritical pressure. The pressure and main steam temperature are 8.5 MPa, 500 C for Unit 1, and 7.3 MPa, 501 C for Unit 2. Superheated steam is made by heating saturated steam obtained at subcritical pressure. Superheated steam (subcritical pressure superheated steam) and supercritical "steam" (supercritical pressure "steam") are different (see Fig. 16.1). The Bonus reactor in the United States was also a subcritical pressure, a test reactor for superheated steam. Superheated steam power reactors were not commercialized in the United States, and have not been used in Russia since. Superheated steam reactors require steam-water separation and recirculation, which is different from the once-through flow type supercritical pressure light water reactor.

C: The supercritical pressure light water-cooled reactor has a compact reactor vessel and main circulation system, so the cooling water inventory is small. Therefore, the heat capacity during loss of feedwater flow or major break loss of coolant accident is small, and the maximum cladding temperature becomes very high.

A: The supercritical pressure light water reactor is a once-through-flow type, so the water inventory inside the reactor pressure vessel is not important. In the once-through-flow type, during a coolant loss accident, coolant flows out from the break point, and a cooling flow is generated in the core and cooled. In contrast, in a pressurized water reactor, coolant flows out from both sides of the break point due to guillotine break of the pipe, and no cooling flow is generated in the core. Even in a boiling water reactor, the core boils due to pipe breakage, but no cooling flow is generated from the core inlet to the outlet. Since the primary cooling system of the PWR includes a steam generator in a closed circulation loop, the outflow rate is maximum at 200% break, but in the once-through-flow type, it is maximum at 100% break.

The safety indicators for supercritical pressure light water reactors are not water levels (water inventory), but flow rates. In the event of an accident, flow rates can be measured more accurately than water levels. The misreading of the water level gauge was the cause of the operators stopping the Emergency Core Cooling System (ECCS) in the TMI accident. In supercritical pressure light water reactors, loss of coolant accidents is not as important as in light water reactors. Safety systems are designed, and safety analyses are conducted to confirm that including the cladding temperature, it is below the criteria. Safety criteria can be met by adding equipment. How simple the safety system can be is important.

C: Because the cooling water is hot, it is difficult to overcome the stress problem of the reactor pressure vessel.

A: The wall of the reactor pressure vessel is cooled by the inlet cooling water (280 C). The same is true for pressurized water reactors. The temperature of that cooling water is also the same. Thermal stress can be alleviated by attaching a sleeve to the outlet nozzle.

C: Instability at startup is a problem.

A: This comment may be from an expert who only knows about pressurized water reactors. Including startup, instability analysis is conducted, and it has been shown that it can be started stably. The problem of instability was tested in the early stages of the development of boiling water reactors and has been overcome. Instability at startup can be avoided by choosing the appropriate ratio of power output to flow. The method of analyzing instability in the licensing of boiling water reactors is also established. The change in water density near the pseudocritical temperature at supercritical pressure is smaller than when boiling in a boiling water reactor, making it less likely to cause instability. Not limited to instability, research results are published as papers after peer-review.

C: A vast amount of research and development on materials and water chemistry is required.

A: The once-through-flow supercritical pressure light water reactor treats all the condensate at low temperature and uses it as feedwater, making it easier to deal with material corrosion than light water reactors. In supercritical pressure thermal power generation, all the condensate is treated and used as feedwater, and the same is true for supercritical pressure light water reactors. Light water reactors use a mixture of recirculated water and feedwater from condensate for cooling. The flow rate of recirculated water is several times that of condensate. Since the recirculated water is at high temperature and high pressure, it is difficult to precipitate impurities by returning it to low temperature and treating all of it. Therefore, it is difficult to deal with the corrosion of the internal structures of boiling water reactors and the steam generator tubes of pressurized water reactors. Experience from supercritical pressure thermal power generation can be used for materials and water chemistry of equipment and piping. The material development and water chemistry of supercritical pressure light water reactors need to be conducted under conditions where all the condensate is purified in a through-flow manner.

C: Compared to the Generation III light water reactors that have been improved and used for many years, supercritical pressure light water reactors have fewer suppliers of equipment and parts, at least in the initial stages.

A: It is inappropriate to compare supercritical pressure light water reactors with light water reactors, for which there are design drawings and construction and operation experience. Light water reactors and their improved versions have less investment risk and are an option for new nuclear power plant construction in many countries around the world. However, in the United States, the cost of natural gas power generation has decreased due to the shale gas revolution, and with the spread of renewable energy, even long-used light water reactor power plants have been forced to close. In countries like the United States, there is a high need for supercritical pressure light water reactors, which have the potential to significantly reduce the construction and power generation costs of light water reactors. In terms of technology and parts, supercritical pressure light water reactors are based on supercritical pressure power generation technology and have many common points. Equipment and parts have experience in thermal power generation. Light water reactors were developed based on subcritical pressure power generation technology of the 1950s. Since then, thermal power generation has transitioned to supercritical pressure. It is natural to aim for the development from light water reactors to supercritical pressure light water reactors.

C: If you try to use chemical volume control (a method of reactivity control that reduces the concentration of boric acid mixed with primary cooling water as fuel burns. Because it reduces the number of control rods, it is used in pressurized water reactors) in a supercritical pressure light water reactor, it will be significantly different from a pressurized water reactor.

A: This comment may also be from an expert who only knows about pressurized water reactors. Chemical volume control cannot be used in light water-cooled reactors where the density of the cooling water in the core changes greatly, such as boiling water reactors and supercritical pressure light water reactors, because the void coefficient becomes positive. In supercritical pressure light water reactors, boric acid water is used as backup equipment for the reactor shutdown system (control rods). This is the same as in boiling water reactors.

C: Online fuel exchange is not possible

A: Due to the lengthening of the operating cycle and the shortening of the shutdown period for maintenance and fuel exchange, the operation rate (capacity factor) of light water reactors is high in many countries, over 90%, and online fuel exchange is not necessary in light water reactors. This is also the case with supercritical pressure light water reactors. In CANDU type supercritical pressure light water reactors, online fuel exchange is possible. Online fuel exchange equipment requires high reliability, such as sealing (seals, etc.) against high temperature and high pressure.

References

Baranaev YD, Kirillov PL, Poplavskii VM and Sharapov VN (2004) Supercritical-pressure water nuclear reactors. At Energy 96:345–351. https://link.springer.com/article/10.1023/B:ATEN.0000038101.90666.da. Accessed 12 Aug 2024

Bittermann D et al. (2003) Economic prospects of the HPLWR. In: Proceedings of GENES/ANP2003, Sept 15–19, 2003, Kyoto, Japan, Paper 1003. Atomic Energy Society of Japan

Glebov AP, Klushin AV (2014) Development of supercritical-water-cooled reactors in russia and abroad. At Energy 116:320–329. https://link.springer.com/article/10.1007/s10512-014-9860-x. Accessed 12 Aug 2024

IAEA (2011) Status report 109–High performance light water reactor (HP-LWR), IAEA 2011. https://aris.iaea.org/PDF/HP-LWR.pdf. Accessed 12 Aug 2024

IAEA (2015) Status report—Chinese supercritical water-cooled reactor (CSR1000), IAEA 2015. https://aris.iaea.org/PDF/CSR1000.pdf. Accessed 12 Aug 2024

INEEL (2003) Feasibility study of supercritical light water cooled reactors for electric power production, 2nd annual report, INEEL/EXT-03-01277. http://large.stanford.edu/courses/2017/ph241/blackwood2/docs/ineel-ext-03-01277.pdf. Accessed 12 Aug 2024

INEEL (2005) Feasibility study of supercritical light water cooled reactors for electric power production, Final report, INEEL/EXT-04-02530. https://inldigitallibrary.inl.gov/sites/sti/sti/3028270.pdf. Accessed 12 Aug 2024

Marchaterre JF, Petric M (1960) Review of the status of supercritical water reactor technology, ANL-6202, Argonne National Laboratory. https://www.osti.gov/servlets/purl/4153321. Accessed 12 Aug 2024

Oka Y, Koshizuka S, Ishiwatari Y, Yamaji A (2010) Super Light Water Reactors and Super Fast Reactors, Springer. https://link.springer.com/book/10.1007/978-1-4419-6035-1. Accessed 12 Aug 2024

Oka Y (2011a) Special lecture, Super LWR and super FR R&D, Joint ICTP-IAEA course on science and technology of supercritical water-cooled reactors (SCWRs) International center for theoretical physics, Trieste, Italy, 27 June to 1 July, 2011. https://indico.ictp.it/event/a10196/session/34/contribution/25/material/0/0.pdf. Accessed 12 Aug 2024

Oka Y (2011b) SC19 plant dynamics and control, Joint ICTP-IAEA course on science and technology of supercritical water-cooled reactors (SCWRs) international center for theoretical physics, Trieste, Italy, 27 June to 1 July, 2011. https://akifumi.yamaji.w.waseda.jp/upload/ICTP-IAEA%20coorse%20SCWR%202011%20Oka%20Plant%20dynamics%20and%20control.pdf. Accessed 12 Aug 2024

Oka Y, Mori H (eds) (2014) Supercritical-pressure Light Water Cooled Reactors, Springer 2014. https://link.springer.com/book/10.1007/978-4-431-55025-9. Accessed 12 Aug 2024

Oka Y (2019) References of Super LWR and Super FR studies of the University of Tokyo and Waseda University at Oka-laboratory February 27, 2019. https://akifumi.yamaji.w.waseda.jp/upload/190228%20Oka%20Reference%20of%20Super%20LWR%20and%20Super%20FR%20studies%20of%20the%20University%20of%20Tokyo%20and%20Waseda%20University.pdf. Accessed 12 Aug 2024

Oka Y (2023) Review of SCWR research in Japan. Front Nucl Eng. https://www.frontiersin.org/journals/nuclear-engineering/articles/10.3389/fnuen.2023.1272766/full. Accessed 12 Aug 2024

Sakurai S et al. (2011) Japanese SCWR fuel and core design study. In: The 5th international symposium SCWR (ISSCWR-5) Vancouver, British Columbia, Canada, March 13–16, 2011, P041, Canadian Nuclear Society

Schulenberg T, Leung LKH, Brady D, Oka Y, Yamada K, Bae Y, Willermoz G (2011) Development of supercritical water-cooled reactor (SCWR) through GIF collaboration. IAEA-CN-164-5S06. https://www-pub.iaea.org/MTCD/Publications/PDF/P1500_CD_Web/htm/pdf/topic5/5S06_H.%20Khartabil.pdf. Accessed 12 Aug 2024

Shioiri A et al. (2003) Development of supercritical-water cooled power reactor conducted by a Japanese joint team. GENES4/ANP2003, Sep. 15–19, Kyoto, Japan, Paper 1121

Shoiri A et al (2004) Technical development for practical application of supercritical pressure water-cooled reactor. Energy systems research institute, Fiscal 2003, Fiscal 2004 (in Japanese). https://www.iae.or.jp/KOUBO/innovation/theme/pdf/h12-3-15.pdf. Accessed 12 Aug 2024. https://www.iae.or.jp/KOUBO/innovation/theme/pdf/h12-3-16.pdf. Accessed 12 Aug 2024

Utterback JM (1994) Mastering the Dynamics of Innovation. Harvard Business School Press

Was G et al. (2007) Corrosion and stress corrosion cracking in supercritical water. J Nucl Mater 371(1–3):1767–201

Winkler F, Knodler D, Huber G (1969) Supercritical Heterogeneous Nuclear Reactor of the Pressure Vessel Type. US Patent 3453178. https://patentimages.storage.googleapis.com/06/28/56/364b1c3f85ac18/US3453178.pdf. Accessed 12 Aug 2024

Wu P et al. (2022) A review of existing super critical water reactor concepts, safety analysis codes and safety characteristics. Prog Nucl Energy 153(39):104409. https://www.sciencedirect.com/science/article/pii/S0149197022002839. Accessed 12 Aug 2024

Yamada K et al. (2011) Overview of the Japanese SCWR concept developed under the GIF collaboration. The 5th International Symposium SCWR (ISSCWR-5) Vancouver, British Columbia, Canada, March 13–16, 2011, P031, Canadian Nuclear Society

Yetisir M et al. (2016) Canadian supercritical water cooled reactor core concept and safety features. CNL Nuclear Review 5(2). Canadian Nuclear Laboratories. https://pubs.cnl.ca/doi/10.12943/CNR.2016.00042. Accessed 12 Aug 2024

Chapter 17
Concept of Nuclear Fusion Reactors

Nuclear fusion is a phenomenon where two light atomic nuclei become one, and the difference in their mass is released as energy. Nuclear fusion occurs in stars like the sun, and on Earth, it has been used since before nuclear fission chain reactions were realized, as a reaction that generates neutrons using an accelerator. In the nuclear fusion reactions occurring in the sun, the expansion and scattering of celestial matter due to the high temperature and pressure generated by the reaction are suppressed by the sun's massive gravity. In accelerator-driven nuclear fusion reactions, the input energy to the accelerator to cause the reaction is much greater than the nuclear fusion energy generated, so unless a way is found to reduce the input energy, it is thought that it cannot be used in a fusion reactor.

A fusion reactor is officially called a controlled thermonuclear fusion power reactor. Control refers to controlling the nuclear fusion reaction to occur steadily within a confined space. An uncontrolled nuclear fusion reaction is a hydrogen bomb. The heat in thermonuclear fusion refers to the process of bringing the fuel nuclei to a state of ultra-high temperature (over 100 million degrees), called plasma, to counteract the Coulomb repulsion force of the positively charged nuclei and cause nuclear fusion, by confining it within a space and increasing the collision frequency between atomic nuclei ions.

There are two methods of confinement for controlled thermonuclear fusion: magnetic confinement and inertial confinement. Magnetic confinement is a method of confining plasma with magnetic field lines created by magnets, while inertial confinement, also known as implosion, is a method of causing a nuclear fusion reaction by uniformly applying energy from the surroundings to a fuel sphere with a laser or the like, making the fuel sphere ultra-dense.

Various magnetic confinement methods have been considered. Broadly speaking, there are mainly mirror methods and torus methods. The mirror method involves placing coils at both ends of the magnetic field lines, and when current is passed through the coils, the magnetic field lines are constricted by the coils, and near the coils, the magnetic moment of the charged particles increases in the direction

Y. Oka, *Nuclear Power Reactor Development*,
https://doi.org/10.1007/978-981-96-2731-8_17

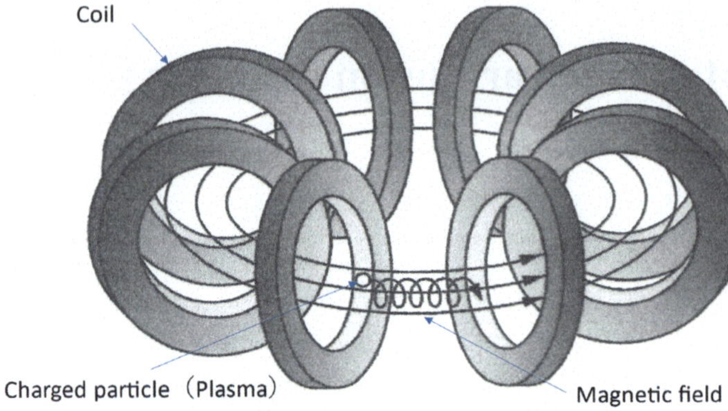

Fig. 17.1 Torus-shaped magnetic field configuration. *Source* Nuclear Handbook, Ohm-sha 2007, pp 917 (in Japanese)

perpendicular to the coil magnetic field, and approaches zero in the direction parallel to the magnetic field, which is used for confinement. The torus method (torus-shaped magnetic field arrangement), for example, as shown in Fig. 17.1, attempts to confine plasma by creating a donut-shaped magnetic field with several coils arranged in a ring, utilizing the fact that charged particles wrap around the magnetic field lines.

In the simple torus method, the internal magnetic field is so strong that the internal plasma expands outward. To prevent this, a twist called a rotational transformation is given to the magnetic field lines. In the torus-type magnetic field arrangement, the major circumference direction is called the toroidal direction, and the minor circumference direction is called the poloidal direction, and giving a rotational transformation means giving a poloidal direction magnetic field component. To give a rotational transformation, there is a method of arranging a helical coil outside and flowing current through it (helical method, external conductor method), and a method of flowing current in the toroidal direction inside the plasma (internal current method). The external conductor method includes stellarators and Heliotrons. The internal current method includes tokamaks and Spheromaks. In addition to these, there is a method called reverse magnetic field pinch (Nuclear Power Handbook 1989 Chapter X Development of Nuclear Fusion, Nuclear Power Handbook 2007 Chapter VIII Research and Development of Nuclear Fusion).

The inertial confinement method proposes a method of directly irradiating a fuel ball with laser light and a method of providing a shell that reflects laser light on the outside of the fuel ball and compressing the fuel ball with laser light reflected by the shell. It is necessary to compress the fuel ball uniformly to high density, but there are issues with uniform compression due to the occurrence of instability. The efficiency of the laser amplifier is also a problem because it is as low as about 1%. In order to achieve break-even where the output energy becomes the same as the input energy to the device due to compression, it is considered necessary to make the ratio of the input energy to the fuel ball to the energy generated by compression more than 100.

In any case, the inertial confinement method has not yet achieved nuclear fusion energy larger than the input energy to the device. In addition, the hydrogen bomb is considered a type of inertial confinement, but since the detonation of the hydrogen bomb uses a nuclear fission chain reaction (atomic bomb), it is premature to think that inertial confinement is realized because there is a hydrogen bomb.

It is sometimes reported that the energy generated by the nuclear fusion reaction is larger than the input energy, but the input energy is not the total energy input of the fusion device. For example, in the case of plasma nuclear fusion, the energy input needs to include the energy for the magnetic field, vacuum, plasma generation, heating device, etc. The input energy for fusion research is, however, the energy incident on the plasma or the target to be compressed. So caution is necessary because it is not the total input energy.

The most advanced research is the tokamak method of magnetic confinement. In Japan, it has been researched for many years at the Nuclear Fusion Research Institute of the Japan Atomic Energy Research Institute (currently the Nuclear Fusion Energy Division of the Quantum Science and Technology Research and Development Organization). The helical method is being researched at the Nuclear Fusion Science Research Institute in Japan. Both are under the jurisdiction of the Ministry of Education, Culture, Sports, Science and Technology. In the United States, under the jurisdiction of the Office of Fusion Energy Sciences (OFES) of the Office of Fusion Energy of the Science Bureau of the Department of Energy, General Atomics in San Diego operates a tokamak nuclear fusion research facility called Doublet III-D (DIII-D) as a joint-use facility, and the Princeton Plasma Physics Laboratory operates a device called National Spherical Torus Experiment-U (NSTX-U) from 1986 and 2016 respectively. The latter is a spherical tokamak device known as a Spheromak. In addition to these, nuclear fusion research is also being conducted at universities such as MIT. The magnetic confinement method has been researched in major European countries. For example, research on confinement by tokamak has been conducted at the Culham Laboratory in the UK with funding from the EU. Internationally, the International Thermonuclear Experimental Reactor (ITER) project, a tokamak method, is being conducted as an international project with the participation of France, Japan, the EU, the United States, Russia, India, and China, and an experimental reactor is being built in Cadarache, southern France. The United States announced the concept of laser nuclear fusion, which had been kept secret until 1972. Inertial fusion is being researched at the Lawrence Livermore National Laboratory (LLNL) under the jurisdiction of the National Nuclear Security Administration (NNSA) of the Department of Energy in the United States. In Japan, research is being conducted at the Laser Science Research Institute of Osaka University.

Nuclear fusion has been researched by excellent physicists around the world for over 50 years using various methods, but it has not yet been successful in generating a large amount of energy that can be used for power generation by stably confining nuclear fusion plasma for a long period of time (for example, several months or more). There have been reports that the energy multiplication ratio (the ratio of generated energy to the heating energy of the plasma) has exceeded 1. ITER aims to achieve long-term operation of several hundred seconds with an energy multiplication ratio

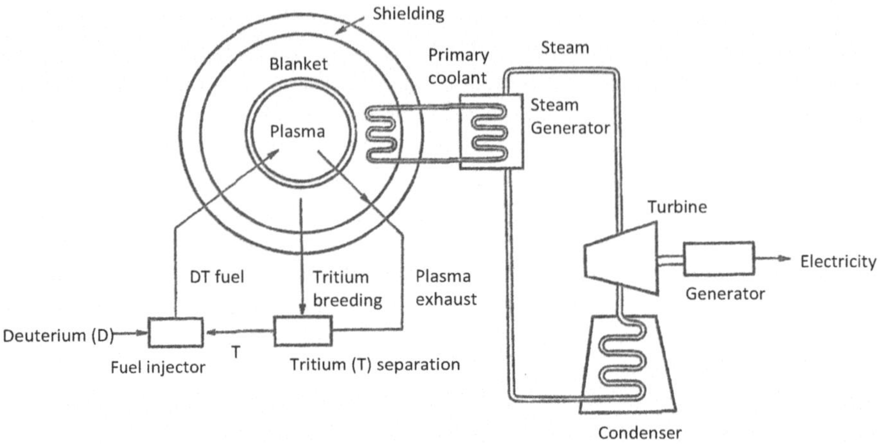

Fig. 17.2 Fusion reactor concept. *Source* Y. Togo and Y. Oka "Introduction of nuclear engineering", Coronasha 1987, pp 142 (In Japanese)

of more than 10, and continuous operation with an energy multiplication ratio of about 5. As already mentioned, the heating energy of the plasma is not all the energy consumed in the nuclear fusion device, so this energy multiplication ratio is not an indicator of the realization of a practical nuclear fusion reactor, but an indicator of magnetic confinement research.

In various nuclear fusion reactions, the reaction with the largest nuclear fusion cross-section (ease of reaction) is the DT reaction, in which tritium (T, tritium) and deuterium (D, deuterium) are fused to produce a helium nucleus and one neutron is emitted. The generated energy is 17.6 MeV, and from the conservation of momentum, the energy is given to helium and neutrons in a 1:4 ratio. In other words, about four-fifths of the nuclear fusion energy (about 14 MeV) is emitted as the energy of the neutron. In nuclear fission reactions, the energy emitted as the kinetic energy of neutrons is about 2.5% of the total, but 80% of the energy from nuclear fusion reactions is released as the neutron's kinetic energy. The kinetic energy of the neutrons is recovered as heat in a part called a blanket that surrounds the nuclear fusion reaction region. The kinetic energy of the charged particles produced by the nuclear fusion reaction (helium in the case of DT reaction) is also given to impurity ions and electromagnetic waves by collision, and becomes the heat of the part called the first wall in front of the blanket.

A fusion reactor is a device that uses the heat generated by nuclear fusion reactions as energy for power generation. Since sustained controlled nuclear fusion reactions have not yet been realized, a fusion reactor is a concept of what such a device should be when it is realized. The conceptual diagram of a fusion reactor is shown in Fig. 17.2.

The plasma that is causing the nuclear fusion reaction is surrounded by the blanket, and radiation shielding is placed on the outside. The blanket recovers nuclear fusion energy and breeds fuel tritium. Since fuel tritium does not exist naturally, a tritium breeding material (a substance containing lithium) is used in the blanket to capture neutrons and generate tritium. This is processed to separate tritium, make DT fuel with deuterium, and inject it into the plasma. The power generation method is the same as the secondary system of the pressurized water light water reactor, and steam is generated to send to the power generation turbine using a steam generator. Although not described in the figure, a toroidal magnetic field coil, a poloidal magnetic field coil and their power supply and cooling system, a vacuum vessel, a plasma heating device, a power supply system, etc. are required.

The concept of a fusion power reactor has been announced many times since the 1970s. In the tokamak type, the Argonne National Laboratory in the United States announced a concept called STARFIRE in 1980. In 1991, based on the knowledge gained up to that time, the Japan Atomic Energy Research Institute announced the concept of a steady-state tokamak power reactor called SSTR (JAERI 1991).

In SSTR, economic considerations are also being made, and the construction cost is 720Gyen (720 billion yen), which is about twice that of a light water reactor. 34.1% of the construction cost is the cost of the tokamak part, and in that, the cost of the toroidal magnetic field coil and the poloidal magnetic field coil is large at 55% combined. The cost of the blanket and vacuum vessel is about 15% of the cost of the tokamak part (about 5% of the total capital cost). The construction cost of the neutral beam injection device and heating device for plasma heating and tokamak current maintenance is 8.3% of the total. The interest during construction is estimated to be low at 8.8% (JAERI 1991, p 596). The construction period is set at 5.5 years. This report states that the SSTR, which is not much different in size from ITER, showed that it can output about 1000MW of electricity with an overall thermal efficiency of 30%.

The economic evaluation can greatly vary depending on its intent, the data used, and the estimation method. It is necessary to disclose the data and methods referred to, but often they are not included in the public report. However, the economic evaluation of SSTR, although conducted by the nuclear fusion research promotion side, is believed to have been conducted with a scientific attitude and has not led to conclusions of low construction costs for the purpose of promoting nuclear fusion reactors. The construction costs of toroidal magnetic field coils, poloidal magnetic field coils, plasma heating systems and their continuous current systems, ultra-low temperature systems, vacuum systems, and their power supply systems are thought to have been evaluated relatively accurately by extrapolating the manufacturing experience of nuclear fusion plasma experimental devices such as JT-60, which were built and started operation at that time. On the other hand, for example, the blanket, which recovers heat from nuclear fusion energy and breeds tritium, does not yet exist as it is designed and made to be used under high-temperature and high-pressure conditions expected in commercial reactors, so the accuracy of its construction cost prediction is not high.

In a nuclear reactor, if a critical amount or more of nuclear fuel is installed in the reactor vessel with a moderator, a nuclear fission chain reaction occurs. There is no need to create and confine a magnetic field like in a nuclear fusion reactor. There is also no need to heat from the outside. Control of the nuclear reactor and sustaining the nuclear fission reaction can be done by moving the neutron absorbing material rod with a motor. From the economic consideration of SSTR, it is understood that a revolutionary innovation is necessary in the method of generating and sustaining nuclear fusion reactions for the practical use (commercial use) of nuclear fusion reactors. In other words, it is difficult to commercialize nuclear fusion on the extension of the current nuclear fusion device. This does not mean that nuclear fusion research is unnecessary. Current nuclear fusion research is in the phase of demonstrating that plasma can be stably confined for a long time. However, for the commercial use of nuclear fusion power reactors, it is considered necessary to find a revolutionary strategy to generate and sustain nuclear fusion reactions. Recognizing the problem correctly is the first step to solving the problem.

Nuclear fusion in the United States has been researched since the 1950s, but the budget increased in the 1970s when there was an oil shock. The Government Accountability Office (GAO) issued a report on nuclear fusion to the federal congress in 1979 (GAO 1979). The report states, "The Department of Energy says that commercial nuclear fusion reactors will be realized from 2025 to 2050, but many challenges remain. Nuclear fusion research is at the level of applied research, and some are still at the level of basic research. Not all issues have been clarified. In budgeting, it should be noted that nuclear fusion is a long-term issue, not a short-term or medium-term issue. The Department of Energy should make a budget proposal based on these considerations."

ITER is one of the international cooperation projects that began as a result of the dialogue between President Reagan and President Gorbachev in 1989. Reports on ITER by the GAO were published in 2007 and 2014 (GAO 2007, 2014).

The 2007 GAO report states, "The U.S. share of ITER is not only the $1.12 billion burden on its construction cost, but also an additional $1.2 billion is needed for operation and decommissioning costs. In order to carry out the construction of ITER on budget and on schedule, it is necessary for the ITER organization to establish a quality control method for the equipment that each country produces in proportion" It recommends that "the progress of inertial fusion research requires the cooperation and talent development strategy of DOE and NNSA".

The 2014 GAO report states, "The construction cost of ITER have expanded and delayed from $1.122 billion in 2005 for seven years (completed in 2013) to $3.915 billion for 28 years (completed in 2034) in 2013, and the U.S. share of $199.5 million in 2014 accounts for 40% of the U.S. fusion research budget. Since ITER is an international project, DOE cannot fully determine the cost and schedule. It can influence, but DOE has not taken formal action for that. The U.S. ITER budget affects the U.S. domestic fusion research budget. In a limited budget, a strategy for the DOE's U.S. fusion program is needed as a reference for congressional budget deliberations. DOE is recommended to clarify the schedule of this international project. DOE agreed with this."

The National Academies of Sciences is an independent, non-profit organization consisting of the National Academy of Sciences, the National Academy of Engineering, the Institute of Medicine, and the National Research Council. The National

Research Council is the working group of the National Academy of Sciences and the National Academy of Engineering. The National Academies of Sciences issued a report in 2019 titled “Final Report of the Committee on a Strategic Plan for U.S. Burning Plasma Research”, recommending that “in order to gain experience in burning plasma at the scale of a power plant, the U.S. should stay in the ITER project in the most cost-effective way and start a national plan for the construction of a compact pilot plant with as low capital costs as possible for power generation by fusion”. Furthermore, at the request of the Department of Energy, the National Academies of Sciences issued “Bridging Fusion to the U.S. Grid” in 2021. The report states, “The United States is a leader in nuclear fusion, and by 2050, the Department of Energy and the private sector should contribute to the transition to low-carbon power sources by generating power on the net from a nuclear fusion pilot plant from 2035 to 2040. In order to commercialize nuclear fusion reactors, the Department of Energy should support the creation of a national team to lead the engineering design and create a roadmap of the concept and technology of the pilot plant, including private partnerships”. The committee of this study is mainly composed of researchers and university professors who have been promoting nuclear fusion research. Since almost no one knows about nuclear fusion research except for nuclear fusion researchers, this might be the case, but this report does not seem to have considered issues of commercialization other than individual technologies, including economics. The GAO mainly discusses procedural issues, and the perspective that the author mentioned with reference to the economic evaluation of SSTR is not seen in the GAO report.

In scientific research, it is necessary to spend a budget and create experimental equipment for the pursuit of truth, but in the field of practical use, commercialization cannot be achieved just by using a budget to create a test reactor, not limited to nuclear fusion.

Not limited to the nuclear field such as nuclear fusion and nuclear fission, it is said that there are three barriers that must be overcome in the process from research and development to practical use in order to realize innovation based on technology: “Devil River”, “Valley of Death”, and “Darwinian Sea”. The “Devil River” is a barrier between the research stage and the product development stage, and in order not to end the research just as research, it is necessary to have the wisdom to link the technology seeds to market needs and conceive specific products. The “Valley of Death” is a barrier between the development stage and the business stage, and in order to connect the product to sales, it is necessary to secure management resources such as funds and personnel. The “Darwinian Sea” is a barrier between the business stage and the industrialization stage, and in order to succeed in business, it is necessary to build competitive advantage and win the survival competition with many rival companies. The policy of the US Department of Energy may be trying to help overcome the “Valley of Death”, but isn’t nuclear fusion research still not overcoming the “Devil River”? Before creating a test reactor for practical use, isn’t innovation to overcome the “Devil River” first necessary? The above three barriers are general ones related to innovation. In the case of power generation technologies such as nuclear power generation, there are unique constraints and conditions related to the business, such as safety regulations and the form of power transmission networks and power businesses, so it is necessary to consider these measures for the realization of innovation.

The Royal Society of the UK held a meeting titled "Fusion Energy Using Tokamaks: Can Development Be Accelerated?" in 2017, and the content is published in an abstract collection called "Philosophical Transactions of The Royal Society A" (Volume 377, Issue 2141, March 25, 2019). There are many papers by nuclear fusion researchers, but at the end, there is a paper titled "Economic Viability of Nuclear Fusion Energy: The Valley of Death and the Innovation Cycle" (Cardozo 2019).

The paper states, "The extremely large investment in the proof-of-principle phase and the long construction period of the fusion reactor plant are obstacles to achieving an effective innovation cycle. A huge investment is needed to move from the development stage to the commercialization stage. If the demonstration reactor (DEMO) could be made small, simple, and inexpensive, the practical application of the fusion reactor could be accelerated." The 2017 meeting of the Royal Society was held on the premise that a demonstration reactor for a fusion reactor is necessary in response to measures to prevent global warming. However, shouldn't the demonstration reactor be something that should be built, with the prospect of not only technical but also economic feasibility? "Small, cheap, and simple" is qualitative, and we are troubled by the slogan. What will readers learn and think from the transition of various power reactor research and development and their commercial use introduced?

Not limited to fusion reactors, the economic feasibility study for practical application should not be mainly conducted by researchers in that field. Price and cost are different, and the reactor manufacturer, who is a stakeholder in the construction, cannot neutrally examine the economy. Specialized knowledge is required for economic feasibility studies. For example, wouldn't it be possible with the cooperation of a power system research institute and a neutral think tank?

Regarding fusion research, it is noteworthy that the director of the Fusion Science Research Institute in Japan has proposed a paradigm shift that segments the fusion energy challenge into academic themes and generalizes each issue to communicate with other fields (Yoshida 2022).

References

Cardozo NJL (2019) Economic aspects of the deployment of fusion energy: the valley of death and the innovation cycle. Phil Trans R Soc A 377:20170444. https://doi.org/10.1098/rsta.2017.0444. Accessed 12 Aug 2024

GAO (1979) Fusion-a possible option for solving long-term energy problems, GAO, U.S. Government Accountability Office. https://www.gao.gov/assets/emd-79-27.pdf. Accessed 12 Aug 2024

GAO (2007) Fusion energy, definitive cost estimates for U.S. contributions to an international experimental reactor and better coordinated doe research are needed GAO-08-30, U. S. Government Accountability Office. https://www.gao.gov/assets/gao-08-30.pdf. Accessed 12 Aug 2024

GAO (2014) Fusion energy, actions needed to finalize cost and schedule estimates for U.S. contributions to an international experimental reactor, GAO-14-499, U. S. government accountability office. https://www.gao.gov/assets/d14499.pdf. Accessed 12 Aug 2024

JAERI (1991) Concept study of the steady state TOKAMAK reactor (SSTR), JAERI-M 91-081, Fusion reactor system laboratory, Japan atomic energy research institute. https://jopss.jaea.go.jp/pdfdata/JAERI-M-91-081.pdf. Accessed 12 Aug 2024

Nuclear Power Handbook (1989) Ohmsha, Published March 30, 1989 (in Japanese)

Nuclear Power Handbook (2007) Ohmsha, Published November 20, 2007 (in Japanese)

Yoshida Z (2022) New era of fusion science institute. J Plasma Fusion Res 98(8):368–374 (in Japanese). https://www.jspf.or.jp/Journal/PDF_JSPF/jspf2022_08/jspf2022_08-368.pdf. Accessed 12 Aug 2024

Chapter 18
Concept of Fusion-Fission Hybrid Reactors

The concept of a nuclear fission–fusion hybrid reactor involves loading nuclear fission fuel material into the blanket of a fusion reactor, causing nuclear fission with fusion neutrons. It uses nuclear fission energy to increase heat generation and improve the economics of fusion reactors.

80% of the energy generated by the DT fusion reaction, which fuses deuterium (D) and tritium (T), is released as neutron energy. The fusion reactor recovers this energy in the blanket and uses it for power generation. Since neutrons decay exponentially in matter, the heat generation of the blanket is large on the surface of the plasma side (first wall) and decays as it goes deeper. By loading nuclear fuel material into the blanket of the fusion reactor, nuclear fission occurs with fusion neutrons, which can flatten the heat distribution of the blanket and improve output (power) density. The irradiation damage to the structural material by neutrons can also be relatively reduced. Since the blanket of the fusion reactor also has the role of breeding tritium (tritium) of the fuel, it is expected to improve the tritium breeding performance by fission neutrons. In nuclear fission reactions, only about 2.5% of the total is released as neutron kinetic energy, so combining nuclear fission reactions with fusion reactions is rational not only for multiplying generated energy but also for flattening heat distribution and relatively reducing neutron damage to structural materials.

Even if nuclear fuel material is loaded, the blanket is subcritical, so a sustained nuclear fission chain reaction does not occur with fission neutrons alone, but the smaller the subcriticality, the larger the energy multiplication factor (the ratio of total generated energy to fusion neutron energy). By keeping the subcriticality constant throughout the operation period, the output of the blanket can be kept constant for a long period without using control rods. This is because the nuclear raw material (U238) in the blanket is converted to fissile material (Pu239), and the converted fissile material undergoes fission. The types of blankets include fast neutron type (without moderator) and thermal neutron type (with neutron moderator) (Oka 1998; Greenspan 1984). Since fusion reactors have not been realized, hybrid fusion reactors

Y. Oka, *Nuclear Power Reactor Development*,
https://doi.org/10.1007/978-981-96-2731-8_18

are also a hypothetical concept. If the cost of a fusion reactor is high, it is considered that even if it is a hybrid reactor, it is economically inferior to a fission reactor.

References

Greenspan E (1984) Fusion-fission hybrid reactors. In: advances in nuclear science and technology 16:289–515. Springer Boston. https://doi.org/10.1007/978-1-4613-2687-8_4. Accessed 12 Aug 2024

Oka Y, Koshizuka S (1998). Conceptual design study of advanced power reactors. Prog Nucl Energy 32(1/2):163–177. https://www.sciencedirect.com/science/article/abs/pii/S0149197097000140. Accessed 12 Aug 2024

Chapter 19
University Nuclear Engineering Education

Nuclear engineering education at universities is the core of human resource development responsible for the research, development, and utilization of nuclear energy. This chapter introduces the mechanism for improving nuclear engineering education at universities in the United States, and discusses points for improvement in Japan. We will talk about the changes in nuclear education at Japanese universities, the creation of nuclear textbooks, and nuclear engineering education at universities in Europe, Korea, and China.

The U.S. Department of Energy and the American Nuclear Society have created a sourcebook for nuclear engineering education in North America (United States and Canada) (Gilligan 2022). According to this, 39 universities in the United States and Canada are providing nuclear engineering education. The number of students in each university's undergraduate (department) and graduate master's and doctoral programs, the names of faculty members, and their areas of expertise are listed, along with 27 research reactors at universities and their types. There are universities with a large number of doctoral students and universities with a large number of master's students. The former are so-called research universities, with the number of doctoral students, for example, being 93 at the University of California, Berkeley, 91 at Massachusetts Institute of Technology (MIT), 111 at the University of Michigan, and 97 at North Carolina State University. The latter are thought to be universities where many students who aim for companies, not researchers, enroll.

University rankings are famous in Japan for each university by a British newspaper company, but in the United States, there are newspaper companies that announce rankings by specialty field, department, and major. These newspaper company rankings are designed to help prospective graduate students decide on their aspirations, with tuition fees and other costs also listed. Some universities list the cost of acquiring credits for each subject, not annual tuition fees. Among these, there is a ranking of nuclear engineering departments and majors in the United States. The ranking evaluation indicators are also disclosed. In addition, there are university rankings in the United States other than this newspaper company.

Y. Oka, *Nuclear Power Reactor Development*,
https://doi.org/10.1007/978-981-96-2731-8_19

Rankings change as the evaluation indicators change, so they are not absolutely important, but since the rankings in the United States are by department, not by university, each department is placed in a competitive environment, prompting improvements from its faculty. Universities operate collectively by department, which is important. University-wide rankings do not directly encourage individual faculty members to make efforts and improvements because the fields are too broad. In the long term, there are departments that have risen in the rankings in the field of nuclear engineering. This department, about 20 years ago, the then department head created a panel to convey the appeal of nuclear power and radiation use, toured high schools to recruit students, and increased the number of enrollees. Although many nuclear engineering departments in the United States have separate departments for radiation-related matters, this department is a nuclear engineering department that is integrated with radiation-related matters.

Among the ranking evaluation indicators is the Graduate Record Examination (GRE). This is a test required to obtain a master's or doctoral degree in either humanities or sciences in the United States. The GRE is operated by ETS, a US company that handles TOEFL, TOEIC, etc. and is designed to measure the logical thinking, reading comprehension, and analytical, vocabulary, and critical thinking skills necessary for graduate school admission. The test is divided into a General Test that asks for general knowledge and a Subject Test that asks for specialized knowledge, and unless you are majoring in something that requires difficult mathematics, physics, etc. only the score of the General Test is required. In an information society, regardless of the field of expertise, we need people with language skills, mathematical thinking skills, and analytical writing skills, and the author feels that such people are the source of America's competitiveness and accountability. What about Japan? In Japan, there are survey results that there are quite a few children who cannot read textbooks.

Rankings can change depending on the indicators used, so they are not the only absolute measure. This is understood in the United States. However, not only in the field of nuclear engineering, but also in the United States, both universities and faculty members are exposed to rigorous evaluations and are placed in an environment where they can hone their skills. Despite being a competitive society, tenured positions, known as "Tenure", are granted to university professors and associate professors in the United States. This is a unique system given to university faculty in the competitive society of the United States. It is believed to recognize the high value of the specialized knowledge that university faculty accumulate over a long period of time. However, even with tenure, it is impossible to continue working as a professor if you cannot secure research funding.

Not only in the field of nuclear engineering, but also among university faculty in Japan, very few have had teaching experience at universities in Europe and the United States. There are times when the author feels that Japanese university education is gradually becoming lenient and the discipline is loosening, but when you are in Japan, you do not notice it. Based on the presentation materials of a few Japanese nuclear engineering faculty members who have experienced both, the author will first introduce education and research at universities in the United States. Next, the author will introduce the example of the Polytechnic University of Milan in Europe.

In the United States, there is no Japanese-style examination competition until high school graduation, but students study seriously after entering university (Ahn 2013; Miwa 2018). In Canada, the university you can enter is determined by your grades in the last two years of high school, so students who want to enter a famous university study quite seriously (Nagasaki 2018). In both the United States and Canada, students cannot graduate if they do not study. In the United States, the order of course registration is determined. For example, if you cannot earn credits in fluid mechanics, you cannot register for heat transfer engineering. There are no make-up exams. Students study desperately because if they cannot earn credits, they will be left behind and may drop out. In the United States and Canada, a significant number of students do not graduate (Miwa 2018). In the United States, there are Community Colleges in addition to universities. This is a public two-year vocational school, and all applicants can enter if they can prove their English proficiency.

In the United States, many students who enter university gain work experience through internships during the summer vacation and include it in their resumes when they find a job. Within the university, students can relatively flexibly choose their own study program, such as double majors and joint majors. The operation of the university is centered on the undergraduate program, which contrasts with Japanese universities where the operation is centered on the graduate program due to the emphasis on graduate schools. Internships for university students are also popular in Canada. There are times when the author feels that Japanese universities have lost rigorous education due to the emphasis on graduate schools.

In the United States, the Accreditation Board for Engineering and Technology (ABET). Reviews of the ABET are conducted every 7 years. ABET is an accreditation agency for engineering education in the United States, which certifies educational programs of engineering universities on a departmental basis. ABET conducts the evaluation and accreditation of engineering education for obtaining engineering degrees in the United States under the commission of the U.S. government. In the United States, there is a Professional Engineer (PE) system, and when manufacturing industrial products, the stamp of a PE in the field is required on drawings and specifications. One of the requirements for obtaining a PE is to have a Bachelor of Engineering degree accredited by ABET. In other words, if you are not a graduate of an engineering department that has been accredited, you cannot work as a PE. Therefore, it is necessary for engineering universities in the United States to receive an evaluation by ABET for their educational programs. In Canada, evaluations are conducted by the Canadian Engineering Accreditation Board (CEAB).

In Japan, there is a system of professional engineers, but it is not mandatory to have the stamp of a professional engineer on drawings and specifications when making industrial products, so there are limits to the spread of the professional engineer system, and there is no accreditation of the curriculum of engineering departments in Japan. The author has heard complaints from U.S. university faculty who are evaluated, asking why they have to be evaluated by a group of engineers, but in the United States and Canada, the evaluations by ABET and CEAB, which are conducted once every 7 years, should be noted as a mechanism for improving university engineering education. The evaluation results are made public. Organizations that do not have a mechanism to improve by receiving criticism, not limited to universities, will deteriorate. There is also the effect of strengthening the unity of the department by the faculty working together to respond to the evaluation. Some departments, like the Nuclear Engineering Department at the University of Michigan, publish the reports they created for the evaluation.

In ABET, a new system was introduced in 2000, transitioning from the past system that emphasized the consistency of the curriculum itself ("what it should be") to a system that focuses on the quality of students as output. Until then, strict constraints were imposed on the curriculum, and its conformity to the standards was examined, but from the late 1990s, mainly private universities in the United States began to introduce engineering education programs that were different from the traditional ones, such as interdisciplinary and managerial skills, and ABET's standards were revised in response to criticism that they did not correspond to such initiatives.

Under the new standards, it is necessary to set the purpose of the curriculum, arrange consistent lectures, etc. to achieve it, and ensure the consistency of the syllabus between each lecture, and the consistency between the curriculum purpose and each lecture. In Japan, in human resource development and its evaluation, we tend to think about what we should teach, but we should pay attention to the fact that it is changing in this way in the United States. The author feels that the transformation of education is supporting innovation with the development of information technology. At the University of California, Berkeley (UCB), faculty measure the state of students at the beginning of each subject and semester. In education, they are required to incorporate social elements such as ethics, lifelong education, and the environment.

The key outcomes to be achieved by graduate students of the UCB Nuclear Engineering Department are as follows (Ahn 2013):

1. The ability to apply knowledge of mathematics, natural science, and engineering to the analysis of nuclear and other systems.
2. The ability to identify, formulate, and solve nuclear engineering problems.
3. The ability to design integrated systems involving nuclear and other physical processes.
4. The ability to design and perform laboratory experiments to gather data, test theories, and solve problems.
5. The ability to learn and work independently, and to practice leadership and teamwork in and across disciplines.
6. The ability for effective oral, graphic, and written communication.
7. A broad education necessary to understand the social, safety, and environmental consequences of engineering decisions, and to engage thoughtfully in public debate on technological issues.
8. An understanding of professional and ethical responsibility.
9. Knowledge of the importance of, and opportunities for, lifelong learning.

It is shown which of these nine items each course corresponds to. These are used in ABET evaluations.

Student evaluations of classes are conducted. Evaluations are done for each course on the last day of the semester. The instructor leaves the classroom and a staff member in charge enters with the forms. The staff member collects the completed forms from the students, confirms that the instructor has submitted the grading for the course to the academic affairs department, and then opens and tallies them. The results are made known to the university headquarters, the department head, and the individual. Students can also find out through the university's website. The evaluations are used for faculty promotions and evaluations. Questions such as "Was this lecture useful for me?" and "Do I think this lecture is necessary for the university?" are asked.

In the case of graduate students, the situation varies greatly by field in the United States. In the case of engineering, all graduate students receive some financial support from their advisors, departments, faculties, and universities, and do not pay tuition themselves. Necessary living expenses are also supported. It is about $2,700/month (as of 2012). Including tuition and salaries, about $60,000 per student per year is needed for California residents, and excellent students apply to multiple prestigious universities, so it is necessary to offer favorable support conditions to attract excellent students.

The curriculum and programs of graduate schools in the United States do not have a clear distinction between doctoral and master's programs at the time of admission. Graduate students take 3–4 courses per semester, which continues for two years. If you aim for a PhD, you need to declare a Major field and two Minor fields, and you need to get the required number of credits and grades (GPA > 3.5) for each. The Grade Points Average (GPA) is a grade evaluation value calculated from the grades of each subject. You take a written exam (Screening Exam) in the first year. You can take the exam twice, but if you fail, you will end with a master's degree. After earning credits and passing the written exam, you take an oral exam (Qualifying Exam). This also has two chances. If you compile a PhD thesis within two years and get the signature of the reviewing faculty, you can get a PhD. The doctorate requires not only research ability demonstrated in the doctoral thesis, but also a wide range of knowledge, so in the United States, there are graduate students who cannot get a doctorate because they cannot pass the written exam.

In the United States, many PhD holders engage in research as postdocs for 2–3 years after obtaining their degrees. During this period, they need to write as many single-authored papers as possible. Supervising professors also encourage them to write single-authored papers and apply for external funding. Postdocs apply for university faculty positions during this period, but recent examples require at least about 10 lead-authored papers. In Canada, it is not required that the papers be single-authored.

Being a postdoc is one of the career paths to becoming a university faculty member, but it is considered risky to continue for more than 5 years, and most postdocs end up working at research institutions or companies instead of becoming university faculty. In Japan, there are opinions about the "postdoc problem" that it is pitiful to remain a postdoc forever, but this is unique to Japan and is lenient. Postdocs who are judged not suitable for university faculty should find other jobs, which should be better for them. In the author's experience, the postdoc problem in Japan is a problem in the science field, and there were hardly any Japanese postdocs in the engineering field.

At the University of California, Berkeley (UCB), faculty members receive nine months' worth of salary from the state government. This corresponds to the salary for the part directly related to "education", such as lectures. For the remaining three months, they pay themselves from their own funds. For example, if the monthly salary is $10,000, approximately $20,000/month is required. About $50,000 to $60,000/year is needed per student. The overhead imposed by the university is 58% at UCB. It is higher at private universities. Faculty members apply for external funding from the Department of Energy (DOE), Nuclear Regulatory Commission (NRC), National Science Foundation (NSF), foreign research institutions, private companies, etc. Associate professors and assistant professors may become principal investigators and hire professors. For example, the Department of Energy supports university faculty through the Nuclear Energy University Program (NEUP), and the Nuclear Regulatory Commission supports them through curriculum development, scholarships and fellowships, and faculty development.

American universities have an internal system to support applications for external funding. Faculty members usually focus on writing the Statement of Work (SOW), which includes the number of graduate students and postdocs to be hired and the necessary equipment. At the department level that supports at the undergraduate level, they prepare the necessary documents for the application, such as a draft budget, based on this. At the campus level, if it is a public call, they prepare the necessary documents as a university and handle the application procedures. If the application is successful, this department negotiates with the sponsor and conducts contract negotiations. In the case of discretionary contracts, they negotiate with the sponsor based on the university's standard contract and also conclude the contract. All public calls from U.S. federal government agencies such as the Department of Energy (DOE) are digitized.

Acquiring excellent faculty is important for the development of a university. According to Prof. Joonhong Ahn's (Toshihiro Ahn in Japanese) materials (Ahn 2013), the text discusses faculty recruitment and hiring, the evaluation and promotion of assistant professors until they acquire tenure (up to 8 years, if tenure is not acquired by then, it will not be renewed), the review of tenure, preparation for its acquisition, evaluation and review of professors and associate professors after acquisition, career paths after obtaining a doctorate, and the personnel desired by the university's leadership and management team.

Next, nuclear education in universities in Japan is discussed (Kudo 2012, MRI Research Associates, Inc. 2021).

It has been about 60 years since nuclear education began in Japan. From the establishment status of nuclear-related faculties and graduate majors (hereinafter referred to as "departments" and "majors") and the transition of the number of graduates, this period can be divided into the following four phases.

(1) From the 1950s to around 1972, the establishment of nuclear-related departments and majors continued, the number of undergraduate graduates and graduate completers continued to increase, and it remained almost constant until around 1992.
(2) Around 1993, under the influence of the emphasis on graduate schools, which shifted the management body of the university from undergraduate departments to graduate majors, the reorganization and name change of nuclear-related departments began, and in many universities, the name of nuclear was removed, and the curriculum content changed.
(3) The 2000s was a period called the nuclear renaissance, when the construction plans for nuclear power plants worldwide increased, and majors with the name of nuclear were revived and newly established.
(4) Since 2011, due to the impact of the TEPCO Fukushima accident, the number of entrants to undergraduate nuclear-related courses first decreased, followed by a decrease in the number of graduate school entrants. However, by 2020, the number of undergraduate entrants has recovered to the level of 2010. In 2024, as in the 2000s, which was called the nuclear renaissance, the number of applicants for nuclear-related departments is increasing.

In Japan, from mid-1950s to around 1972, nuclear research institutes were newly established at 13 universities (Hokkaido University, Tohoku University, Tsukuba University, University of Tokyo, Tokyo University of Mercantile Marine, Musashi Institute of Technology, Tokai University, Nagoya University, Kyoto University, Osaka University, Kinki University, Kobe University of Mercantile Marine, and Kyushu University) and other departments/majors, as well as at Tokyo Institute of Technology and Rikkyo University. During this period, the writing of nuclear power textbooks for lectures and the translation of American textbooks were actively carried out. However, in 1991, the university establishment standards were revised, and a significant change occurred in undergraduate education. As it is called graduate school departmentalization, the unit of university management shifted from each department to each major in the graduate school. The education of each specialty in each department at the undergraduate level changed to a curriculum that emphasizes the basics of engineering, and many departments were broadly categorized into university departments, and many universities lost departments with the name of nuclear engineering. The name change is not limited to nuclear engineering, and departments closely related to specific specialties and their industries, such as shipbuilding and precision machinery, have been integrated into university departments. Many universities also changed the name of the graduate school major from nuclear engineering to quantum engineering, environmental energy engineering, etc.

The unit of university management shifted from the department of the faculty that conducts education to the major of the graduate school that conducts education and research, and the budget called school expenses allocated to each faculty member from the government increased by about 50% because research is also conducted. This also became a factor in promoting the departmentalization of graduate schools as a university. There are departments (faculties) that conduct education and research institutes that aim for research at universities, but the budget allocated to the department (faculty) in charge of education is much less per faculty member than the budget allocated to the research institute for the establishment and maintenance of research facilities. Faculty members of the graduate school also served as faculty members of the undergraduate school, and research has been conducted in the graduate school with the participation of graduate students from the past. The departmentalization of the graduate school, where the allocation budget from the government increases because research is also conducted, was welcomed by the faculty of the undergraduate school, which is the majority of university faculty.

In addition, since April 2004, national universities in Japan have been incorporated, and each university has become a "National University Corporation" with an independent corporate personality from the administrative organs set up by the Ministry of Education, Culture, Sports, Science and Technology. It is said that the target of administrative reform to reduce the number of national civil servants was directed at national universities, which have a large number of national civil servants, and incorporation was carried out. National university faculty members became de facto civil servants due to incorporation, but even after subsequent administrative reforms, the number of civil servants, including de facto civil servants, has not decreased. Even now, the national budget is allocated to the National University Corporation of Japan. State universities in the United States receive state budgets as faculty salaries, etc. As introduced in the materials of Prof. Toshihiro Ahn (Joonhong Ahn) of the University of California, Berkeley, there may be many points that can be referred to in the operation of the National University Corporation of Japan in state universities in the United States. In Japan, the graduate school has become the center of education and management, but in the United States, the undergraduate school is the center of education and management. The United States is superior.

Innovation in the United States comes far more from universities than from national research institutions. The presence of many venture companies around university campuses is not limited to the United States. Universities play a role in attracting and supplying excellent talent from around the world. There are many points that Japan can learn from the operation and roles of universities in the United States and other major foreign countries. Research funding is necessary, but it is also necessary to understand that simply providing a large budget does not necessarily achieve the goal. Scientific research funding is a mechanism that encourages creativity and ingenuity among university faculty and is important.

In Karlsruhe, Germany, there was the Federal Nuclear Research Center (FZK Karlsruhe), which is now the Karlsruhe Institute of Technology (KIT), formed in 2009 by the merger of the long-established University of Karlsruhe and the Federal Karlsruhe National Research Center, conducting research and education. Its research organization consists of numerous research groups for each theme, aiming for practical application-oriented innovation. It's as if each university laboratory is independent, acquiring the necessary research equipment and attempting innovation and practical application. This is a new form of research university.

In Japan, university faculty members involved in nuclear power have formed a voluntary organization called the University Nuclear Faculty Council and are active, exchanging information at the annual and autumn meetings of the Atomic Energy Society of Japan. Currently, 19 universities are providing nuclear education in some form at the graduate level. Among them, the ones that have a major with the name "nuclear" are the University of Tokyo, Tokyo Institute of Technology, Tokyo City University, Waseda University, Nagaoka University of Technology, University of Fukui, and Kyoto University. Other universities are providing nuclear education under the name of a major that includes words such as "quantum", "energy", and "environment". Kyoto University is the only university that has been providing nuclear education under the name "nuclear" since the establishment of the Department of Nuclear Engineering in 1957.

The University of Tokyo once changed its name to the Department of System Quantum Engineering, but in 2005, based on the faculty quota of the Nuclear Engineering Research Facility affiliated with the Graduate School of Engineering, which was operated in Tokai Village, Ibaraki Prefecture, a small number of faculty members transferred from the Department of System Quantum Engineering participated, and Department of Nuclear Engineering and Management and the Nuclear Professional School were established. With the establishment, the Nuclear Engineering Research Laboratory and the Comprehensive Nuclear Research Center of the University of Tokyo were abolished. The Nuclear Professional School is located in Tokai-mura, Ibaraki Prefecture. The Department of Nuclear Engineering and Management is located on the Hongo Campus in Tokyo.

The University of Tokyo's Nuclear Professional School is a one-year master's program primarily for students employed by companies and government organizations, located in Tokai Village, Ibaraki Prefecture. This major was established when the Japan Atomic Energy Research Institute considered creating a nuclear education school, but realized that it would be difficult to do so without being a university that can issue "degrees". In the Nuclear Professional School, education is conducted with the cooperation of faculty members from the University of Tokyo, researchers from the Japan Atomic Energy Agency, and lecturers from nuclear power plant manufacturing companies. Many of the students are working adults, such as employees of electric power companies and officials of government agencies related to nuclear power. Graduates who have completed the designated group of subjects with grades above the standard are exempted from the written examination for the national qualification of Chief Reactor Engineer. Therefore, grade certification is strict, and there have been students in the past who were unable to obtain this qualification. In the Nuclear Power Major, strict nuclear power education (credit certification) equivalent to that of American universities is conducted. In the nuclear professional school, students do not conduct research and are not required to write a master's thesis. The faculty of the Professional Nuclear School, like when they were at the Nuclear Engineering Research Laboratory, also handle the Nuclear Engineering and Management Major, and graduate students of the Major are conducting research in these faculty's laboratories. The author led the establishment of the school and the Major.

In the case of the author, the motivation for establishing the Nuclear Professional School was not necessarily working adult education, but the need to realize strict nuclear engineering education equivalent to that in the United States in Japan.

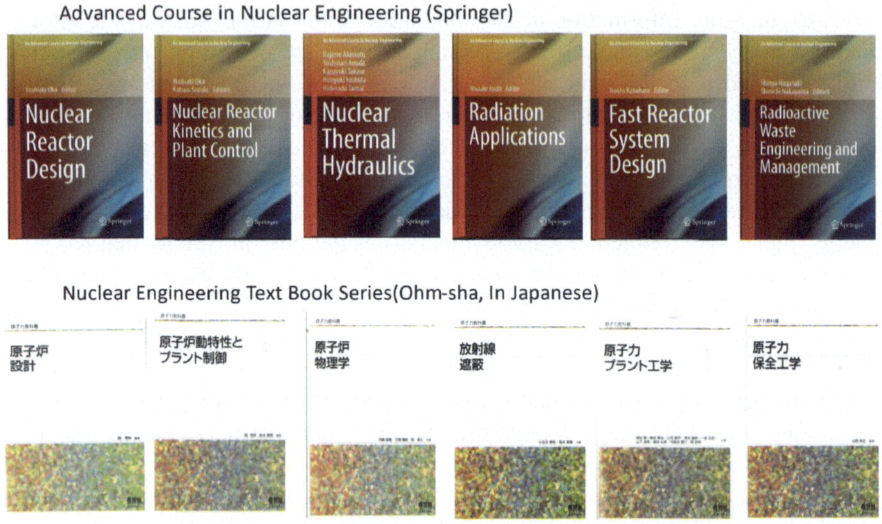

Fig. 19.1 Nuclear engineering textbooks

Personnel who have acquired a solid system of knowledge are essential for the use of nuclear power and its development. For several years after its establishment, the lectures of the Nuclear Professional School were online, and they could be audited at the Nuclear Engineering and Management Major in Hongo campus in Tokyo, and credits could be earned, but it seems that this is not currently being done. The author retired in 2010 and reflects that the sharing of significance and purpose was insufficient.

Based on the education of the Nuclear Professional School, a series of nuclear engineering textbooks was created with the cooperation of the lecturers, researchers, and instructors who gave the lectures. More than 10 nuclear power textbooks have been published (Fig. 19.1). These reflect 40 years of experience in the use and research and development of nuclear power in Japan. English versions have also been created and published from Springer.

Another motivation for the establishment of Nuclear Engineering and Management Major and Nuclear Professional School at the University of Tokyo was to apply the competitive budget for a large-scale education program that the Ministry of Education, Culture, Sports, Science and Technology was promoting at the time. The University of Tokyo's nuclear engineering research facility was a research institute, not an education institute, so it could not apply for the 21st Century COE Program, which was a hot topic at the time. By creating the Nuclear Education and Management Major, we were able to apply for the Global COE Program following the 21st Century COE program, and in 2007, the "Nuclear Education Research Initiative Leading the World" was adopted. The program was implemented in collaboration with the Graduate School of Public Policy of the University of Tokyo's Law School and the Department of Nuclear Engineering at the University of California, Berkeley.

The Cooperative Major in Nuclear Energy of Waseda University and Tokyo City University is a joint graduate school where faculty members from both universities share the lectures of a single curriculum. This major was established in 2010 as a result of the introduction of a joint education program in the university establishment standards in Japan, which allows multiple universities to jointly realize attractive education and human resource development. The presidents of both universities led the establishment. Classes are held at the Shibuya Satellite Campus of Tokyo City University. Entrance examinations are conducted at each university, and students are assigned to the research laboratories of the faculty members of the university they passed. The master's thesis review is conducted with the participation of faculty members from both universities. In addition to the Cooperative Major in Nuclear Energy, Waseda University established two other joint graduate schools (majors) in 2010. The author was actively involved in the establishment and operation of Cooperative Major in Nuclear Energy of the Waseda University after retired from the University of Tokyo in 2010.

Fukui Prefecture is home to many nuclear power plants. Fukui University established the affiliated International Nuclear Engineering Research Institute in Tsuruga City in 2009. With four fields and seven departments, it is a large research institute for a university's nuclear research institute. The same year, they established a Nuclear and Safety Engineering Sub-specialization Course in the Faculty of Engineering. In 2011, they established a new Nuclear and Energy Safety Engineering Specialization in the Graduate School of Engineering, and have been intensifying nuclear education by the institute's faculty.

There are five research reactors in Japan for university use, built at Kinki University, Rikkyo University, Musashi Institute of Technology (now Tokyo City University), Kyoto University, and the University of Tokyo. Four of these were put into operation in 1960s. Each university established a nuclear research institute to manage operations and conduct educational research. Currently, two reactors at Kinki University and Kyoto University are in operation, but the Kyoto University Reactor (KUR) is scheduled to cease operation in 2026. The Kyoto University's Institute for Integrated Radiation and Nuclear Science (formerly Kyoto University Research Reactor Institute, KURRI) is a national joint-use research institute, and the KUR is available for joint use by universities nationwide. Research reactors at other universities also accepted use by other universities. The reason for the cessation of KUR operation is explained as the deadline for the United States to take back spent fuel and the aging of the facility. The spent fuel from the reactors at Rikkyo University, Musashi Institute of Technology, and the University of Tokyo has also been handed over to the United States. The fact that the United States was able to take it back was fortunate for the universities in order to proceed with the decommissioning measures of the research reactors. Kyoto University has a device called a critical assembly, which has extremely low output and changes the core configuration to conduct critical experiments, and will continue to be used for educational research in combination with an accelerator.

Research reactors come in various types and outputs. Taking advantage of their characteristics, various research were conducted not only for education and training but also for analysis using generated neutrons, material research, medical applications, neutron measurement, material testing, etc. The Japan Atomic Energy Research Institute (now JAEA) has built eight research reactors, test reactors, and experimental reactors, some of which have already transitioned to decommissioning measures, and as of 2024, four are in operation or scheduled to be in operation. Two research reactors were also built by nuclear power plant manufacturing companies (heavy electric companies) in the 1960s, but they are in the process of decommissioning. As the number of research reactors in Japan has decreased, plans to build a test research reactor on the site of the fast prototype reactor "Monju" in Tsuruga City, Fukui Prefecture, are being considered.

The author has been involved in research using the University of Tokyo's fast neutron source reactor "Yayoi", joint use, and operation management for a long period of time, so the author will describe that experience. This reactor is a research reactor of a rare fast reactor in the world that maintains a nuclear fission chain reaction with fast neutrons, and was operated from 1971 to 2011. The thermal output is 2kW. Using the generated fast neutrons, research was conducted on fast neutron transport and shielding, pulse-shaped supercritical operation characteristic tests, development of epi-thermal neutron irradiation fields for boron neutron capture therapy, development of fast neutron measurement and standard fields, neutron damage to materials, tritium breeding performance which is the fuel of a fusion reactor, and radiation decomposition of water by fast neutrons (water chemistry). Off-pile research (research not using a reactor) included design research, heat transfer fluid dynamics, various calculation methods, and research on radioactive waste behavior (Oka 1998b).

The University of Tokyo Reactor was built adjacent to the Tokai Research Institute of the Japan Atomic Energy Research Institute (now the Japan Atomic Energy Agency) in Tokai Village, Ibaraki Prefecture, and a nuclear research facility was established at the Faculty of Engineering, University of Tokyo, where its operation and management were carried out. Following the construction of the University of Tokyo Reactor, a linear electron accelerator and basic experimental devices for fusion reactor blanket design were built there. An ion accelerator, known as a heavy irradiation research facility, was also set up. The linear electron accelerator generates the world's shortest electron pulses, and research on radiation chemistry, a basic process of chemical reactions caused by radiation, was conducted along with the development of short electron pulse generation methods. In the basic experimental device for fusion reactor blanket design, electromagnetic structural mechanics, which is the structural mechanics using magnetic fields such as fusion reactors and accelerators, was developed. It is fracture mechanics of volumetric force loads. By combining an ion accelerator and an electron microscope for in situ observation studies of irradiation damage were conducted. All of these are world-leading research results.

In university organizations, educational departments (faculties and graduate schools) and research departments (research institutes) are separate departments in Japan. The Nuclear Engineering Research Laboratory of the University of Tokyo was an affiliated facility of the educational department, so its faculty members also served in departments and graduate school majors, and many undergraduate and graduate students were able to use the facility of Nuclear Engineering Research Laboratory for research with the faculty. Generally, the affiliated facilities of the faculty are small in scale, do not have large equipment like the research reactor and accelerators, and have few faculty members. Nuclear Engineering Research Laboratory had a research institute-scale number of faculty members (4 departments) and facilities, and many undergraduate and graduate students were able to conduct research for their graduation theses, master's theses, and doctoral dissertations using distinctive large-scale equipment such as the University of Tokyo Reactor and the linear electron accelerator. The environment where many students and graduate students can conduct research using large experimental measures is rare in Japanese universities. It was fortunate that there was continuous support for the research facilities from the Ministry of Education, Culture, Sports, Science and Technology, which is in charge. It is clear that this method is excellent in university education and research, as the above-mentioned world-class research was conducted with the involvement of many excellent graduate students.

The University of Tokyo's reactor ended its life as a research device and ceased operation in 2011. To be precise, it was scheduled to stop at the end of March 2011, but it automatically stopped due to the major earthquake on March 11, 2011, and did not operate thereafter. Another reason for the cessation of operation was that regulations on reactor management gradually became stricter in Japan. In particular, compared to the time of construction, the requirements for fuel management and security have gradually become stricter. Technical management staff were provided, but because it was affiliated to educational department, the number of staff was smaller compared to research institutes, and the burden on technical staff to respond to increased regulatory requirements increased. The author investigated and summarized the regulation and management of research reactors at American universities, but there was no improvement effect. The reason why the Japan Atomic Energy Agency proposed the establishment of Nuclear Professional School to the University of Tokyo seems to be not only because it cannot issue degrees, but also because it could not overlook the plight of the facility management of the University of Tokyo's reactor. At that time, the Ministry of Education and the Science and Technology Agency were integrated, and it was thought that the barriers to the use of research facilities of the Japan Atomic Energy Agency would be lowered, which is why the signboard of Nuclear Engineering Research Laboratory was changed to Nuclear Professional School. Looking at the subsequent progress, the expectation that the barriers would be lowered was not met. Regarding research and development facilities and their maintenance and improvement, it is not limited to the nuclear field, but when comparing then and now, the impact of the Japanese government's financial strain may be significant. In any case, an environment where graduate students can experiment using excellent research facilities is important along with the cultivation of good instructors in the future.

After graduating from universities and other institutions, the Nuclear Human Resource Development Center of the Japan Atomic Energy Agency has been active since the dawn of nuclear use in Japan. It conducts short-term and medium-term (up to two months) training courses for radiation and reactor technicians, cooperation through dispatch of experts to universities, etc. and international training for neighboring Asian countries. As an industry activity, the Japan Atomic Industrial Forum (JAIF) has created a nuclear human resource development network and is working with universities and international organizations to promote nuclear human resource development. As an academic activity, each specialized division of the Atomic Energy Society of Japan holds a summer seminar and conducts specialized training.

There are not many published materials about their own human resource development in research and development institutions and companies, but there are presentation materials from those who have been operating research institutes for many years at nuclear reactor manufacturing companies (Kiguchi 2017). This material states that the first step in human resource development is to set the desired human resource image, which should be reviewed according to needs, and that the key to human resource development is the supervisor, and the basic is on-the-job training (OJT). The Atomic Energy Commission has summarized its views on human resource development in the nuclear field (Japan Atomic Energy Commission 2018).

Next, we discuss nuclear education in European universities. Nuclear education in European universities is described on the homepage of the European Nuclear Society, including undergraduate, master's, doctoral programs, and even the European Nuclear Energy Network (ENEN, explained in the ENEN + Project section (ENS 2024)).

Nuclear education is conducted in many universities in Europe. In undergraduate programs, nuclear education is conducted within departments such as physics, energy engineering, and power engineering. In graduate schools, universities with nuclear engineering are common in countries with nuclear power plants. Doctoral programs often bear the name of nuclear physics or power engineering, but there are also universities that offer degrees in nuclear engineering. Compared to the United States, there are fewer universities with departments or majors that only teach nuclear engineering. Universities with a large number of faculty members in nuclear engineering include the Polytechnic University of Milan in Italy (Politecnico di Milano) and Imperial College London in the UK. Other than these, there are the Royal Institute of Technology in Sweden, the University of Pisa, and others.

Italy abolished nuclear power plants after the Chernobyl accident, but nuclear education at universities has been conducted at several universities. At the Department of Nuclear Engineering of the Polytechnic University of Milan, the largest and most systematic nuclear engineering education in Europe is being conducted (Ninokata 2019).

The Department of Energy Engineering at the Polytechnic University of Milan (Politecnico di Milan), specifically the nuclear power sector, has many faculty members in six groups, including nuclear reactors, nuclear physics, and radiation measurement. Students who have studied basic engineering such as physics and chemistry are studying nuclear engineering in the master's program of graduate school. All lectures are conducted in English. Education is the fundamental idea, and the department checks the content of the lectures, and there is an obligation to conduct regular interviews with students, and it is said that the faculty members have no time for research. 70 to 80 percent of the entrants complete the program within the prescribed two years. The university has invested more than 2.5 billion yen (approximately 2M Euro) and completed the Nuclear Engineering Experimental Building in 2015. Despite the TEPCO Fukushima accident, the attitude of the Italian government and the university to invest in nuclear education and research has not changed. Furthermore, there is a research implementation company called SIET, which is famous in the world's nuclear industry, that conducts large-scale tests of nuclear reactor components and systems in the suburbs, and the operation management of the university's experimental group is outsourced to this company. The number of students is increasing, the rate of advancement to doctoral programs is high, and the employment situation is extremely good, and it is said that there are many graduates from the Polytechnic University of Milan among the faculty of famous universities in the United States. The nuclear engineering course at the Polytechnic University of Milan is supported by students and the Italian Ministry of Education. Future efforts include acquiring competitive funds within Europe, acquiring international students, and establishing a double degree program with universities in China, a new nuclear power country. In Europe, the Erasmus plan, which aims to improve the quality of education by promoting exchanges between higher education institutions in European countries, has been promoted, and the European credit transfer system ETCS was established in 1989. In the case of the Polytechnic University of Milan, the learning amount of 1 ECTS is defined as the learning outcome that a typical student is expected to acquire in 25 h of study.

At the Imperial College London in the UK, faculty members in fields related to nuclear power and radiation, such as mechanical engineering, chemical engineering, civil and environmental engineering, materials engineering, and earth science and engineering, have formed a center and are operating a graduate education program in nuclear engineering.

France is a major nuclear power country, but famous universities in France are smaller than those in the United States and the UK, and there is no large-scale nuclear engineering department in France, but nuclear education is conducted at the University of Paris-Sud, ParisTech, Ecole Centrale Paris (ECP)—Supelec and the French Alternative Energy and Atomic Energy Commission (CEA)'s ISTN, etc. (I2EN 2018). ParisTech is composed of several science and engineering universities in and around Paris, including the École Polytechnique and the Paris School of Mines, which are the pinnacle of French science and engineering. Many of the executives and administrators in France's nuclear-related ministries, power companies, and corporations are graduates of the Paris School of Mines and École Polytechnique. At École Polytechnique, students from major universities in Europe, China, the Middle East, and African countries are studying physics and mathematics through double degree programs and study abroad systems. CEA's ISTN offers various training courses in nuclear technology in cooperation with French universities. At ISTN of CEA Cadarache Institute, they provide education in reactor physics and training in operation of pressurized water reactors and business skills. CEA accepts many graduate students and allows them to conduct doctoral thesis research.

In Germany, education and research in nuclear engineering were conducted in departments such as mechanical engineering. The Federal Research Institute in Karlsruhe, which was conducting nuclear research, has transformed into a technical university, where education and research on nuclear power are also conducted. There are many professors in fields other than nuclear power, each of whom has created their own research institute and conducts research while supervising graduate students.

In South Korea, the Seoul National University, Korea Advance Institute of Science and Technology (KAIST), and Hanyang University have departments of nuclear engineering, where nuclear engineering education was conducted for many years. Since the 2000s, which was called the nuclear renaissance, the number of universities conducting nuclear education has increased. South Korea has a nuclear research institute called KAERI, where many doctoral graduates are conducting research.

In China, nuclear engineering education has been conducted at Tsinghua University, Shanghai Jiaotong University, and Xi'an Jiaotong University since ancient times. With the explosive expansion of nuclear power utilization since 2000, the number of Chinese universities conducting nuclear engineering education has greatly increased. For the training of technicians for rapidly expanding nuclear power plants, faculty members of university departments of nuclear engineering and power engineering cooperated as instructors in training seminars. Tsinghua University has a large technical research institute, Institute of Nuclear and New Energy Technology (INET), which was established in 1960 and developed a high-temperature gas-cooled reactor. In China's government-affiliated technology development organizations, there is Nuclear Power Institute of China (NPIC) in Chengdu, Sichuan Province, which is conducting technology development and testing of light water reactors and others.

References

Ahn J (2013) Current status of education and research at U.S. Universities, presentation material of Waseda university cooperative major in nuclear energy (in Japanese). https://akifumi.yamaji.w.waseda.jp/upload/UC.pdf. Accessed 12 Aug 2024

ENS (2024) Education, European nuclear society. https://www.euronuclear.org/education-training/education/. Accessed 12 Aug 2024

Gilligan J, Baily S (eds) (2022) Nuclear science & engineering education sourcebook 2022, North American Edition, version 6.22, American nuclear society, U.S. Department of Energy. https://neup.inl.gov/SiteAssets/FY2022_Documents/Nuclear%20Science%20%26%20Engineering%20Education%20Sourcebook%202022.pdf. Accessed 12 Aug 2024

I2EN (2018) French nuclear education and training, support to newcomer and expanding countries, international institute of nuclear energy (I2EN). https://www.i2en.fr/wp-content/uploads/2018/09/BROCHURE2018_I2EN_compressed-1.pdf. Accessed 12 Aug 2024

Japan Atomic Energy Commission (2018) On human resource development in the nuclear field (Opinion) (in Japanese). https://www.aec.go.jp/kettei/seimei/kenkai/20180227.pdf. Accessed 12 Aug 2024

Kiguchi T (2017) About human resource development (My Opinion), 3rd atomic energy commission document no. 1 (in Japanese). https://www.aec.go.jp/kaigi/teirei/2017/siryo03/1_haifu.pdf. Accessed 12 Aug 2024

Kudo K (2012) Current status of nuclear education in universities. J AtIc Energy Soc Jpn 54(11):707–710 (in Japanese). https://www.jstage.jst.go.jp/article/jaesjb/54/11/54_707/_pdf/-char/ja. Accessed 12 Aug 2024

Miwa S (2018) Nuclear education at Purdue University, 32nd Nuclear Committee Document No.1 (in Japanese). https://www.aec.go.jp/kaigi/teirei/2018/siryo32/1_haifu.pdf. Accessed 12 Aug 2024

MRI Research Associates, Inc. (2021) Basic survey for securing and developing human resources in the nuclear field in the future report, commissioned report by the Cabinet Office (in Japanese). https://wwwa.cao.go.jp/oaep/dl/search_houkoku2021.pdf. Accessed 12 Aug 2024

Nagasaki S (2018) Nuclear education at McMaster University and Others. 36th Atomic Energy Commission Document No.1 (in Japanese). https://www.aec.go.jp/kaigi/teirei/2018/siryo36/1_haifu.pdf. Accessed 12 Aug 2024

Ninokata H (2019) About nuclear education at Politecnico di Milano in Italy, (in Japanese). 12th Atomic Energy Commission Document No.1. https://www.aec.go.jp/kaigi/teirei/2019/siryo12/1_haifu.pdf. Accessed 12 Aug 2024

Oka Y (1998) Fast neutron source reactor and related nuclear engineering research of the university of Tokyo-a comprehensive review, Special Issue. Prog Nucl Energy (1/2). https://www.sciencedirect.com/journal/progress-in-nuclear-energy/vol/32/issue/1. Accessed 12 Aug 2024

Chapter 20
Research and Development and Knowledge Base for Nuclear Utilization

A knowledge base is necessary for the use of nuclear power. The knowledge base refers to the accumulated and systematized knowledge for nuclear utilization, the personnel who can acquire and create new knowledge, and the experimental and research equipment for research and development. Research and development play a role in creating a knowledge base for nuclear power.

While it is essential to receive systematic education in specialized fields at universities and acquire cultivation as a prerequisite, the methods to build a knowledge base include.

(1) Cultivating personnel by participating in research and development,
(2) Creating and publishing reports summarizing the knowledge and experience gained through research and development,
(3) Develop calculation codes incorporating knowledge gained from research and development to improve predictability,
(4) Hold seminars and workshops,
(5) Hold summer schools for young people,
(6) Hold result presentations and symposiums,
(7) Hold training sessions for experts,
(8) Create and publish research papers, reports, reviews, and explanations, etc.

The basis for these activities is the existence of research and development equipment and information libraries at universities and research and development institutions, and the presence of leaders and research groups in laboratories, etc.

The budget supporting these research and development activities includes facility maintenance costs and researcher salaries for research and development institutions, faculty salaries and research funds for universities, and various competitive funds for research and development and human resource development. In the case of nuclear engineering, the U.S. Department of Energy and the Nuclear Regulatory Commission provide competitive funds and scholarships for universities and research and development institutions. In Europe, in addition to the budget provided by the country

Y. Oka, *Nuclear Power Reactor Development*,
https://doi.org/10.1007/978-981-96-2731-8_20

for research and education at research and development institutions and universities, there are research and human resource development programs provided by the European Community, and research and education are conducted in cooperation with various European countries. This is not limited to the nuclear field. In Japan, the Ministry of Education, Culture, Sports, Science and Technology and the Nuclear Regulation Committee/Nuclear Regulation Agency provide research funds and competitive funds for research and development institutions and universities through their operating organizations. For university faculty, the Grants-in-Aid for Scientific Research from the Japan Society for the Promotion of Science are important research funds.

In graduate school, there is a saying that education and research are one. The research conducted by graduate students for their master's and doctoral theses is education through research. Knowledge and experience can be gained by thinking and researching on one's own. In the United States, part of the research projects acquired by university faculty is undertaken by graduate students and postdocs, who conduct research while receiving a salary at universities and national research institutes. This allows graduate students to gain experience not only in research methods, but also in the management of research and development and research projects, and to improve their abilities by compiling reports.

In Europe, in addition to conducting research in each country's nuclear research program, researchers from various European countries have been conducting research in teams in the European Community's programs and OECD/NEA research projects. The European Community (EU) started a multi-year research grant program called the Framework Program (FP) in 1984, targeting EU member countries. The research period was initially three years, but it became four years in FP4 from 1994, and seven years in FP7 from 2007. The budget also increased significantly. FP8 started in 2014 and was called Horizon 2020. From 2021, it is called Horizon. The research plan called FP9 in Europe is being implemented in various fields (JST 2021).

Examples of knowledge base construction through research and development of nuclear reactor safety research, such as core meltdown accidents of nuclear reactors, are described for Europe and the United States. Chapter 5 introduces the results in the research and development of severe accidents, where in Europe, 51 universities and research and development institutions participated in severe accident research called SARNET in the FP6 framework program (Albiol 2010). As a result, severe accident research, which was scattered in each European country, began to be conducted in collaboration and cooperation, calculation methods and evaluation methods were standardized, and calculation codes for analyzing the behavior of severe accidents at nuclear power plants were also created. The research results were stored in a database in a common format. A common method for probabilistic risk assessment was developed. A book (Sehgal 2012) and short-term educational courses on severe accidents were created. This book summarizes the findings of severe accident research and is compiled with the cooperation of SARNET by a professor who has been conducting severe accident research at the Royal Institute of Technology in Sweden for many years. Human exchanges between research and development organizations in Europe have also deepened. These have built a knowledge base in the field of severe accidents.

The OECD/NEA has had a committee (CSNI) from its inception where member countries gather to discuss safety issues and have been implementing joint experimental and analytical projects. Thermal fluid dynamics, calculation code development and verification, nuclear fuel behavior and cladding development, and severe accident research were conducted in cooperation with member countries. They created analytical models, verified calculation codes, and created reports that have been useful for systematizing and inheriting knowledge. Test facilities such as fuel test reactors and hot cells owned by member countries have been utilized. Since severe accident experiments handle high-temperature radioactive materials, the facilities are expensive, so they were conducted by sharing themes through international cooperation. The experimental results have also been useful in the development of accident response methods (NEA 2012).

The Cadarache Research Institute of France's CEA and the Karlsruhe Research Institute of Germany (which merged with the Karlsruhe Institute of Technology in 2009) have been conducting severe accident research using research facilities funded by their respective governments, as well as implementing the OECD/NEA's severe accident research project. For example, numerous studies have been conducted, such as France's research on the behavior of nuclear fission materials during accidents (PHEBUS), and Germany's research on the release behavior of nuclear fission materials from overheated fuel rods (QUENCH). In the NEA's project, graduate students also participate in research presentations, where they have the opportunity to hear the latest results and receive comments on their own research from researchers. Reports have been compiled to understand the behavior of severe accidents in light water reactors, not just to enhance the understanding of individual phenomena, and calculation codes have been verified (IRSN 2007). Through these, the knowledge gained from research and development is systematized, predictability is improved, and the knowledge gained is passed on and human resources are developed.

The U.S. Nuclear Regulatory Commission has conducted international cooperative research on severe accidents, including the Cooperative Severe Accident Research Program (CSARP). This research began in 1988, and international cooperation on severe accident phenomena and calculation code development was conducted with the participation of U.S. national research institutes, French and German research and development institutions, and the OECD/NEA. In the United States, experiments such as the interaction between core melt and concrete were conducted.

In the United States, the Safety Research Group of Sandia National Laboratories has been conducting safety research and severe accident research with the support of the Nuclear Regulatory Commission. The knowledge gained is compiled into the severe accident calculation code MELCOR, which is used to improve understanding and predictability of severe accidents. Furthermore, this research group has compiled a report on light water reactor safety (NRC 2002). Although not public, they have also created training materials for Nuclear Regulatory Commission staff. Through these activities, a knowledge base for nuclear reactor safety has been built. The U.S. nuclear industry has considered severe accidents from the perspective of accident management and created the calculation code MAAP. MAAP was created to serve as a simulator for accident management, and its purpose of development is different from MELCOR, which is a calculation code that follows the physical behavior of severe accident phenomena. There are also books written from the perspective of severe accident management (Henry 2011).

What about Japan? Is a knowledge base like that of the West being built? Numerous research papers have been published. However, in the field of severe accidents and nuclear reactor safety, no literature equivalent to the Western reports and reviews introduced here can be found, even with a search. There are also virtually no self-made calculation codes that systematize knowledge, especially for severe accident analysis. There are books that describe safety in relation to probabilistic risk assessment and regulatory rules, but there are none that comprehensively summarize severe accident phenomena and safety. Japan is partial and fragmented. Safety research has, however, been conducted for many years in both industry and research and development institutions.

Not limited to the field of safety, the Japan Atomic Energy Agency, which represents Japan's nuclear power, has produced various reports, among which are a group of reports classified as reviews. However, most of them are annual reports and activity reports. No report that systematically organizes knowledge in a certain field from an overview perspective can be found. Furthermore, there is no summary that summarizes the main points in the report. It is written in the preface what the report was made for. In Western reports, you can grasp the main points by reading the initial summary. If you want to know the details, you can deepen your understanding by reading the description in the text afterward. Isn't it that the work of creating an overview, systematic report is not being done in the management of researchers' work? The basic thing of writing a summary in the report is also forgotten. This is an organizational management issue. The summary needs to be created with the same level of effort and attention as the main text. Not only report creation, but also human resource development and systematizing knowledge in calculation codes are necessary. Isn't Japan's knowledge base for nuclear power use extremely fragile compared to the West?

Japan's nuclear disaster prevention guidelines have procedures written, but not the thinking. To write the thinking, systematic knowledge and knowledge in related fields are needed, which can be obtained from a group of reports that describe systematically and comprehensively and from human resource development. It is not enough for researchers to think from the bottom up and create research papers in subdivided fields. It is necessary to let them research and work so that they can systematize the information obtained as knowledge. Western research and development programs are created in this way, and organizational management is also conducted. These are necessary for Japan to promote robust nuclear power use. The Atomic Energy Commission has pointed out the importance of building a knowledge base and has also described it in the White Paper on Atomic Energy (Japan Atomic Energy Commission 2020). Although it is described in comparison with the West here, those who have been involved in nuclear reactor safety for many years in domestic companies also point out problems in slightly different expressions (Sawada 2015). The reason why improvements do not progress even if problems are pointed out is an organizational management issue, which leads to the problem that the audit system is not functioning in Japan.

References

Albiol T et al (2010) SARNET: severe accident research network of excellence. Prog Nucl Energy 52(1):2–10. https://www.sciencedirect.com/science/article/abs/pii/S0149197009001504. Accessed 12Aug 2024

Henry RE (2011) TMI-2: an event in accident management for light water moderated reactors. American Nuclear Society

IRSN (2007) Research and development with regard to severe accidents in pressurised water reactors: summary and outlook, Report IRSN-2007/83. https://inis.iaea.org/collection/NCLCollectionStore/_Public/43/012/43012266.pdf?r=1&r=1. Accessed 12 Aug 2024

JST (2021) Overseas survey report, EU research and innovation framework program horizon Europe, CRDS-FY2021-OR-02, Research and Development Center, Japan Science and Technology Agency (in Japanese). https://www.jst.go.jp/crds/pdf/2021/OR/CRDS-FY2021-OR-02.pdf. Accessed 12 Aug 2024

Japan Atomic Energy Commission (2020). Chapter 8: strengthening the foundation of nuclear power utilization, 2019 Edition of the atomic energy white paper, atomic energy commission (in Japanese). https://www.aec.go.jp/kettei/hakusho/2019/. Accessed 12 Aug 2024

NEA (2012). Main benefits from 30 years of joint projects in nuclear safety, OECD/NEA, NEA No. 7073. https://www.oecd-nea.org/jcms/pl_14784/main-benefits-from-30-years-of-joint-projects-in-nuclear-safety?details=true. Accessed 12 Aug 2024

NRC (2002) Perspectives on reactor safety, NUREG/CR-6042 Rev2, SAND93-0971, U.S. Nuclear regulatory commission. https://www.nrc.gov/reading-rm/doc-collections/nuregs/contract/cr6042/index.html. Accessed 12 Aug 2024

Sawada T (2015) The way of nuclear reactor safety research, trends in academia, issue, pp 60–63. Japan academic cooperation foundation (in Japanese). https://www.jstage.jst.go.jp/article/tits/20/2/20_2_60/_pdf/-char/ja. Accessed 12 Aug 2024

Sehgal BR (2012) Nuclear safety in light water reactors: severe accident phenomenology. Academic Press

Chapter 21
Summary and Discussion

First, we need to learn from past lessons about research and development. This chapter explains the differences in power reactor development methods in each country to date, and summarizes the lessons learned and future challenges. For the practical application of research and development, it is necessary to understand not only the technical aspects but also the issues of interaction with society.

Globally, light water reactors dominate the market for power generation reactors. The main reasons are that light water is the most efficient and compact moderator compared to graphite and heavy water, and it has a high commonality with thermal power generation technology. The former can be explained from the theory of nuclear reactors (the theory of neutron moderation in nuclear reactor physics). Light water reactors require low enriched uranium as fuel, but there is an international competitive market for low enriched uranium, and it is easy for private companies to obtain it at a reasonable price.

Graphite has poor moderation efficiency and a large core volume, resulting in high capital costs and high power generation costs. The graphite-moderated carbon dioxide-cooled reactor (160 MWe) of Japan Atomic Power Co., Ltd. in Tokai Village has almost the same size as the adjacent boiling water light water reactor (1100 MWe). The power generation cost of the graphite-moderated reactor was also high. In terms of decommissioning measures, graphite cannot be processed and disposed of as easily as light water. The UK has been operating its independently developed graphite-moderated carbon dioxide-cooled power reactor for many years, but is transitioning to light water reactors. In the UK, in the past, there was a proliferation of reactor manufacturers, and the influence of national research and development institutions and state-owned power companies on the practical application of power reactors was significant, and light water reactor manufacturers did not grow.

Heavy water, although not as much as graphite, has poor moderation efficiency compared to light water, and the core volume is large (power density is low), so it cannot be said to be a good moderator. Natural uranium can be used as fuel,

Y. Oka, *Nuclear Power Reactor Development*,
https://doi.org/10.1007/978-981-96-2731-8_21

but a heavy water production facility is required. The advantage of a heavy water-moderated power reactor is that it is a pressure tube type, so it does not require large equipment that can be manufactured in limited factories, like a nuclear reactor pressure vessel. Canada exports CANDU reactors to the world, and India uses them for power generation after domestic production.

Sodium-cooled fast reactors are considerably more expensive than light water reactors due to the unique plant systems and equipment required for liquid sodium coolant. Fast reactors require a reprocessing plant and a MOX fuel processing plant to obtain fuel, necessitating their development and commercialization. Since there is no global competitive market for these, they are a unique type of reactor for commercial power generation. There is a competitive market for the enriched uranium fuel of light water reactors. France is at the forefront of the world in fuel cycle businesses such as reprocessing, but power companies have suffered significant losses from the commercial reactor of sodium-cooled fast reactors built in the past. Researchers at nuclear research and development institutions hope to conduct research and development, but the multi-year energy plan of France in 2018 hardly mentions fast reactors. The United States was promoting the construction of the Clinch River Fast Breeder Prototype Reactor by both Congress and the responsible agencies, but the construction cost soared and the private power companies did not bear the additional burden, so the development budget was cut off in 1983 and the plan was canceled. In this case, whether private investment is made worked as a check function for the practical application of the new reactor.

Researchers are engaged in research and development because they believe the type of reactor they are studying is necessary. Researchers like the type of reactor they are studying. Technical information on a specific reactor type is held by the group of experts involved in that research, and is scarce among other experts. Research and development institutions need to secure research and development funds. For these reasons, there is a lot of information from the promotion side in industry magazines, academic journals, their news, etc. If you ask researchers or those involved in research institutions about the necessity of their research and development, it is natural that the answer will be that it is necessary. This situation is not limited to power reactors and the nuclear field.

There are hardly any reports summarizing past research and development experiences from perspectives other than technology, based on technical knowledge. The fact that information in the field is biased toward technical information may be the same in other engineering fields. Nuclear power generation is an energy policy, and dependence on the country is also large. For these reasons, it tends to be difficult to correct the course of nuclear research and development. This book summarizes past experiences with references. The author hopes this book will be used to pave the way for the future of power reactors. The same mistakes and similar mistakes need to be prevented.

The practical model of a new type of nuclear reactor, developed by national research and development institutions and used by the private sector, was the development model for Japan's power generation nuclear reactors. However, there are no engineers capable of manufacturing design in research and development institutions, and there are no factories to make equipment. Therefore, even if a research institution develops it, the design drawings are made and manufactured by private companies that are nuclear reactor manufacturers. Nuclear reactor manufacturers are grateful for orders, so they are cooperative with the development policy. In Japan, the national policy was determined by being dominated by information from the pro-nuclear side such as research and development institutions and nuclear reactor manufacturers, and the check function against the administration did not work. Even if they developed and made prototype reactors called "Fugen" and "Monju", they were not commercialized as commercial reactors. In the meantime, at "Monju" and "Joyo", the knowledge and technical capabilities of these nuclear reactor types of nuclear reactor manufacturers have decreased from the time of construction, and they have been criticized for failures. Light water reactors developed by the private sector in the United States were also made in large numbers by electric power companies in Japan. New type conversion reactors and fast reactors could not win the competition with them. Japanese power companies are private companies, and the fact that investment in the construction of commercial reactors of these reactor types developed by the country has not been made can be considered as a check function working there. However, if there was a mechanism to correct at an early stage, or another development model in Japan, it should have been possible to effectively use the excellent human resources and budget that were involved in the development.

In the early days when power generation nuclear reactors were developed in the world (1950s to 1960s), government agencies developed power generation nuclear reactors. It was in the UK, France, Canada, etc. In the United States, private companies developed light water reactors. The U.S. government cooperated in the test by making a test site at a national research institute. Currently, in Europe and the United States, the government agencies do not develop power generation nuclear reactors themselves. In these countries, the role of nuclear research and development institutions is not the practical application of power generation nuclear reactors, but the construction of a knowledge base related to nuclear power utilization, that is, the discovery, accumulation, and systematization of knowledge for nuclear power utilization through research, and the development of calculation codes. In Japan, the role of the Japan Atomic Energy Agency (especially the former Power Reactor and Nuclear Fuel Development Cooperation side), which is defined by laws and regulations, is old, and it was pointed out that it should be reviewed when the author was the chairman of the Atomic Energy Commission, but it was not understood. The construction and operation of a test reactor (experimental reactor) of a new type of reactor is considered to be difficult for a private company, which is a nuclear reactor manufacturer alone, so it should be the role of a national research and development institution, but if a small test reactor is made in the future, some kind of ingenuity that makes use of past experience and reflection is necessary. Space development has been carried out with the involvement of the country, along with nuclear development. Now private companies are entering and it is international competition. The only way to go to space is a rocket, but there are means other than nuclear power for power generation, so the competition is more intense. The goal of supplying electricity cheaply and stably is important.

In the United States, there is a system in which the government supports the cost of building prototype reactors by venture companies. This system (a model from development to commercialization) is different from the model of Japan's power reactor development, where the government develops and the private sector uses it. In Japan, national research and development institutions developed new reactors on the premise that they would be used in the private sector, but they have not actually been used for commercial purposes. If we had learned from precedents and lessons in Europe and the United States, we could have made course corrections. In order to build 10 new reactors, it is necessary to have an international market in mind from the beginning. It should be noted that not all new reactors by venture companies are successful. If we do not learn from past history and lessons, the cost and effort will be wasted. The author hopes this book will be useful for that purpose.

The use and development models of nuclear power generation vary from country to country. Japan's nuclear research and development has been done with reference to the United States, so the author will talk about the United States. The United States commercialized the pressurized water light water reactor developed by the Navy by enlarging it by private companies. The boiling water light water reactor was developed by GE, a large private company in the United States. Although the test reactors and experimental reactors were built at the Idaho National Laboratory, a national facility, GE has consistently avoided government involvement, such as cooperation with national research institutes, due to intellectual property rights such as patents. In the United States, many national research institutes were conducting nuclear research and development in the early days, but reforms were made in the 1990s to make the Idaho National Laboratory the core research institute for nuclear research and development. The national research institutes in the United States have not been responsible for practical application (construction and operation of prototype reactors and their commercialization) since before, and have been conducting research and development in role sharing with private companies. After that, the government supported the use of nuclear power generation and its research and development in the NP2010 plan aiming for the construction of new light water reactors, the NERI plan for research and development of Generation IV nuclear reactors, and so on.

Since the late 2000s, in the United States and other countries, venture companies have been working toward the new construction of small reactors (nuclear reactors with an electric output of 300MW or less) and modular reactors (nuclear power plants with a modular design with many small output reactors lined up). In this regard, in September 2022, the U.S. Government Accountability Office (GAO) issued a report stating that an independent external review is necessary to prevent misuse of support costs, reveal potential organizational biases that have led to project failures and cost increases in the past, and prevent systemic failures for the three new reactor demonstration projects, including the construction of experimental reactors supported by the U.S. Department of Energy (DOE), totaling $4.6 billion. The Department of Energy has announced that it will follow this recommendation.

The U.S. Department of Energy's new reactor demonstration project is a mechanism to encourage innovation in nuclear reactors by venture companies, but it is costly because it immediately supports the construction of experimental reactors. When the government supports, as was the case in the UK in the past, the private companies supported will be multiple. In the U.S., three projects have been selected, totaling $4.6 billion. What the U.S. will do is the responsibility of the U.S., but first, it is necessary to learn from the lessons of past development. On top of that, there is room to devise cost-effective methods for the innovation of power reactors. For example, currently, the concept of a new reactor can be created and its performance can be confirmed by analysis with a calculation code. The development of calculation codes is something that national research and development institutions are good at. It is sufficient to improve the existing calculation code by adding the calculation model necessary for the new reactor. This is possible if venture companies cooperate with national research and development institutions. The cost required at this stage is probably about 1/100 of the cost of building an experimental reactor. Make multiple new reactors compete, evaluate the results, and narrow down to a few. Furthermore, after that, the structural design of the experimental reactor and the consideration of extrapolation to commercial reactors are carried out, and the experimental reactor to be built is selected by evaluating the quantity, economy, and reliability. The calculation code created in the first stage can be used for approval and subsequent research and development. Since a large number of people and budgets are required for equipment structural design, it is considered to be done with cost sharing between venture companies, power companies, and the government. The results of the structural design can be utilized in the construction of experimental reactors. The cost required at this stage is probably about 1/10 of the cost of building an experimental reactor. The evaluation of the results should be done by a third-party organization, and the results should be published in a document. This is an example, but it would be good to consider what kind of model is good. In that case, wouldn't an independent check mechanism for policies, like the audit offices and administrative inspection offices in Europe and the United States, be necessary?

In the United States, the GAO and the federal congress's administrative check mechanism are functioning not only for nuclear power policy. Japan also has an audit office, but its role is to inspect accounting fraud, and it does not function as a mechanism to check the results of administration like the U.S. GAO. The Japanese Audit Office Law is different from the United States. The GAO and the audit office do not state the pros and cons of policies, but have the role of clarifying the facts based on their investigative authority and reporting them. Japanese administrative officials are excellent, but an administrative check mechanism is necessary. Without a check mechanism, no matter how excellent, it is impossible to respond to environmental changes and proceed while effectively using taxes. Relatedly, referring to the United States, a short and easy-to-understand explanation of policies for the public by administrative officials (section chiefs) is required for Japanese ministries to improve the accountability of the responsible administration.

In comparison with the United States, another point to mention about Japan in terms of research and development is that while research and development reports funded by the government are very detailed in the United States, they are quite simple in Japan. If we compare the references in this chapter, the second- and third-year reports of the INEEL's research on supercritical pressure light water reactors each have about 150 pages, and the research results can be understood in detail. In addition to this, short reports are produced every quarter. The annual budget and monthly usage are also posted as figures. The contents and results of the sub-themes handled by WH Corporation are also described in the report. With these, the obtained research results can be handed over as knowledge for future reference. The report of Japan's nuclear reactor manufacturer is less than 50 pages, and there is no detailed report that is publicly available, and it is only once. The research conducted by this manufacturer under the cooperation of the Generation IV International Forum (GIF) has probably not been published except for individual international conference papers, presumably due to a small budget. The results of research and development conducted using tax money, even if conducted by private companies, need to be reported and made public, and the knowledge gained needs to be passed on to future generations. This leads to future development. It is difficult to understand this necessity from a closed perspective within a company or a country. The research cost of INEEL's SCWR is about $300,000 per year for three themes (about $100,000 each), which is not a large research and development budget in the United States. This is not about the example of the supercritical pressure light water reactor, but detailed reports are also being produced in Japan, but they are not public and may be lost over time. The knowledge gained can only be utilized by sharing it. The results of research and development conducted using tax money should be made public.

In the United States, national research institutes and others make past research and development reports public, and they can be obtained on the Internet. The public information of the Nuclear Regulatory Commission (NRC) includes research and development reports in addition to regulations and safety matters, and is substantial.

The United States has a public Internet library called the UNT Digital Library. This is an Internet version of the information of American university libraries, aimed at providing and preserving information related to academia. Not only university research and doctoral theses, but also government and other web information, such as the aforementioned INEEL reports, are included and made public in the research and development reports of national research institutes. It is not only used by university students, but also helps people around the world learn throughout their lives. Furthermore, in the United States, there is a digital library called HathiTrust Digital Library, which digitally archives books managed by universities and other libraries and provides them to university students, contributing to the inheritance of knowledge. Creation, publication, and sharing of information are essential for innovation in an information society. Japan is lagging behind in this respect. The use of Artificial Intelligence (AI) has become the third boom. The success or failure of AI use may depend on the quality of the data used.

If we talk about the reports on past nuclear power-related research and development in Japan, the reports on research and development conducted by the Nuclear Power Engineering Corporation (NUPEC, established in 1976, transferred most of its operations to the Japan Nuclear Energy Safety Organization (JNES) in 2003, which was integrated with the Nuclear Regulation Authority (NRA) in 2014) are not left in a form that can be made public. Many research and development projects, such as those related to earthquake resistance and heat transfer fluid dynamics, were conducted at the Earthquake Testing Center in Tadotsu city under their jurisdiction or commissioned to nuclear reactor manufacturers. It may have contributed to the improvement of the technical capabilities of nuclear reactor manufacturers, but it is necessary to disclose the research and development conducted using tax money, which should have thickened the foundation of knowledge about nuclear power generation, inherited the information obtained, and contributed to future innovation, robust use of nuclear power, its global deployment, regulations, and safety assurance.

The knowledge that a single person can experience in a lifetime is limited. If the knowledge gained is not written down, other people cannot gain that knowledge or experience without having the same experience. As time passes, the knowledge and experience existing in society degrade. It is essential for improving competitiveness in an information society to write down and disclose information on research and development, etc. conducted using tax money, and make it available for social use. In these respects, isn't Japan significantly inferior to the United States?

Adding to the necessity of creating and disclosing research and development information, information related to regulations, including the information on the underlying research and development, should be disclosed. Without enhancing the disclosed information along with the establishment of the cost–benefit approach in regulations, it is impossible to conduct rational regulations.

Nuclear power utilization requires a knowledge base; a knowledge base is a report summarizing the results of research and development, calculation codes, personnel who have acquired knowledge, and experimental equipment and research equipment for generating knowledge. Japan has many research papers and reports on finely divided themes, but there are hardly any reports that collect and summarize the results of research and development. There are few reports whose abstracts are created so that the key points of the content can be understood. This is not limited to the fields of safety and severe accidents, which were explained as examples. The knowledge base for Japan's use of nuclear power is extremely fragile and needs to be improved. It is not always necessary to repeat the research conducted in the past in Europe and the United States. The information is in English reports, so it is conceivable to collect, organize, and summarize them and share them organizationally. In the information age, machine translation has developed and AI technology has advanced, so there should be new methods different from the past.

There are many areas in which university education in Japan should be improved compared to the United States and Europe. In Japan, both students and parents believe that once you enter university, it's a given that you will graduate. Even in Japanese universities, if there are compulsory subjects, you will have to repeat a year if you cannot earn credits for those subjects. However, since the emphasis on graduate schools, university education has become more lenient, the number of compulsory subjects has decreased, and there are fewer faculty members who strictly certify credits. In American universities, the order of courses is specified, and for example, you cannot take a course in heat transfer fluid dynamics without earning credits in thermodynamics. Compared to the United States and Canada, university education in Japan, not limited to nuclear power, has become lax. The lack of a mechanism to encourage improvement in universities and faculty, or the fact that it has been neutered, is common with other areas in Japan such as administration and research and development. In American and Canadian universities, the improvement mechanism is functioning. The educational evaluation conducted to give the professional engineer qualification to graduates is helping to improve education in the United States and Canada. Student course evaluations are academically shown to be highly biased, but they are reflected in faculty evaluations. Faculty members not only lecture in the classroom, but also spend time helping students improve their studies and understanding.

Japanese university faculty members are forced to spend a lot of time on administrative work. Rules should be simplified and transparent to reduce the burden of faculty work. It is a negative for university management to have high-paid faculty members do clerical work. Many Japanese faculty members have to hire secretaries with limited research funds for the operation of budget management in their laboratories. In American universities, a secretary, who is a regular university employee in each department, does all the administrative work. They also prepare for information exchange meetings. Accounting is just submitting receipts to the university headquarters. The budget is no single-year accounting, so there is no need to use it all up, and it can be carried over to the next year. Simplification of university administrative rules is important for the internationalization of faculty. The author has hired foreigners as full-time lecturers two times. It was difficult to keep hiring them because they could not understand the procedure in Japanese and could not share the administrative work of the laboratory. If you can't speak Japanese, you can't be a university faculty member in Japan, and Japanese universities will fall behind the world. The most important thing for a university is to hire excellent faculty. Chores can be reduced by simplifying and making rules transparent. There is no concept of chores for American university faculty.

Recently, Tohoku University has been highly evaluated in Japan, but when the author was promoting exchanges with the Department of Nuclear Engineering at the University of California, Berkeley, with funds from the Ministry of Education, Culture, Sports, Science and Technology's Global COE around 2007, Tohoku University was exchanging with Berkeley as a whole university. The author speculates that the effect of this has become apparent. There are many points that can be referenced for the education, research, and operation of Japanese national universities in the state universities in the US.

This book discusses the history of research and development of power reactors, but in the field of nuclear power, not only power reactors, but also many research and development activities are being conducted. The Atomic Energy Society of Japan has established the Atomic Energy Historic Award in 2008 to commemorate its 50th anniversary, and has been honoring its achievements. The first time, it honored many achievements up to that point (Atomic Energy Society of Japan 2008). Through these, we can learn about Japan's nuclear research and development. The author was fortunate to be able to serve as the president of the society at this 50th anniversary. The

American Nuclear Society has established the Nuclear Historic Landmark award to honor achievements. Through this, we can learn about a part of the historical nuclear research and development in the United States (ANS 2021).

To add a final word, "The application of similar technologies from different fields is the key to new discoveries and inventions". This is the word of a friend who has been engaged in research and development in the automobile industry for many years. In other words, "Research should not be vertical, but horizontal". Vertical refers to digging deep within a certain field. Of course, mastering specialized knowledge in a certain field is essential.

References

ANS (2021) Nuclear historic landmark award, Award recipients, Honors and Awards, American nuclear society. https://www.ans.org/honors/recipients/award-nuclandmark/. Accessed 12 Aug 2024

Atomic Energy Society of Japan (2008) 1st (Fiscal 2008) Atomic energy historic award, atomic energy society of Japan (in Japanese). https://www.aesj.net/about_us/awards/1st-kouchikusyo. Accessed 12 Aug 2024

Chapter 22
Exercises

Exercises are provided below for better understanding.

1. Summarize the history of the development and commercialization of pressurized water reactors in the United States, and improvements in technology.
2. Summarize the history of the development and commercialization of boiling water reactors in the United States, and improvements in technology.
3. Summarize the methods and history of power enhancement of light water reactors in the United States.
4. Summarize the improvements in light water reactor fuel.
5. Summarize the development and commercial use of graphite-moderated gas-cooled power reactors in the UK.
6. Summarize the development of sodium-cooled fast breeder reactors in the United States, the UK, France, Germany, and Russia.
7. Summarize the development of high-temperature gas reactors in the United States and Europe.
8. Summarize the development, use, and export of light water reactors and RBMKs in the former Soviet Union and Russia.
9. Summarize the development, use, and export of light water reactors in France.
10. Summarize the development and export of power reactors including light water reactors in Germany.
11. Summarize the development and use of power reactors including light water reactors in Japan.
12. Summarize the use of light water reactors in Sweden and the transition of nuclear power policy.
13. Summarize the history and current situation of the use of light water reactors in Finland.
14. Summarize the use of power reactors in South Korea, and the research, development, and export of light water reactors.
15. Summarize the development, utilization, and export of China's power generation nuclear reactors.

Y. Oka, *Nuclear Power Reactor Development*,
https://doi.org/10.1007/978-981-96-2731-8_22

16. Summarize the development, utilization, and export of Canada's CANDU reactors.
17. Summarize the U.S. policy to promote the construction of light water reactors since the 1990s and its results, also mention about the export (no need to include plans).
18. Summarize the history of research for the use of thorium resources for power generation in the United States.
19. Summarize the price changes of natural uranium ore since the 1960s.
20. Summarize the use of plutonium in light water reactors in Europe. Summarize the changes in the amount of plutonium held by countries that have conducted plutonium utilization in light water reactors.
21. Summarize the arguments and issues of the U.S. Carter administration and the subsequent U.S. non-proliferation group regarding reprocessing policy.
22. Summarize the history and current status of plasma fusion research in the United States.
23. Summarize the history and current status of plasma fusion research in Japan.
24. Summarize the history and current status of research on severe accidents (core meltdown accidents) of light water reactors in the United States.
25. Summarize the history and current status of research on severe accidents (core meltdown accidents) of light water reactors in Europe (France, Germany, European Union, etc.).
26. Summarize the development and features of the U.S. light water reactor core meltdown accident analysis codes MELCOR and MAAP.
27. Summarize the results of Japan's nuclear reactor safety research and severe accident research by searching for published literature. Compare and discuss these results and contents with the results and contents of nuclear reactor safety research and severe accident research in the United States and Europe.

Afterword

Various types of reactors have been developed in Japan. Not only the pressurized water reactors and boiling water reactors that are in practical use, but also the sodium-cooled fast reactors "Joyo" and "Monju", the pressure tube type heavy water-moderated boiling light water-cooled reactor "Fugen", and the high-temperature gas-cooled test reactor (HTTR). The nuclear-powered ship "Mutsu" was also developed. The Calder Hall type graphite-moderated carbon dioxide-cooled reactor, the first commercial power plant in Japan, was imported from the UK for use. Japan is a rare country in the world that has construction and operation experience of power reactor types such as pressurized water reactors, boiling water reactors, sodium-cooled fast reactors, heavy water moderated reactors, graphite-moderated reactors, and high-temperature gas cooled reactors.

The author would like to talk about his experience in nuclear reactor design and analysis. After completing the doctoral program, the author took a job in the Reactor Design Engineering Research Group of the University of Tokyo's Nuclear Engineering Research Laboratory in Tokai Village, Ibaraki Prefecture. Since then, the author has specialized in nuclear reactor design study and led an academic life. At the university, the author lectured on nuclear reactor physics, nuclear reactor kinetics, and nuclear reactor design, and studied and used heat transfer fluid dynamics and safety in the research. The author conducted a seminar on probabilistic risk assessment for the industry once a year for 25 years. After becoming a professor, the author ran his research group with an associate professor specializing in heat transfer fluid dynamics and advanced research on nuclear reactor design. In addition, in the research group, we developed methods such as the MPS method for analyzing solid and fluid with calculation points and the sensitivity and uncertainty analysis method for neutron transport calculations, so our research results were not limited to nuclear reactor design.

Even though the author specializes in nuclear reactor design, the university is not a nuclear reactor manufacturing company, so the author doesn't have experience in manufacturing design of nuclear power plant equipment, but he has examined

Y. Oka, *Nuclear Power Reactor Development*,
https://doi.org/10.1007/978-981-96-2731-8

various nuclear reactor concepts. Even if he didn't research it himself, the laboratory which the author belonged to when young was conducting research on new types of reactors such as fast breeder reactors, so he had many opportunities to learn about the concept examination and analysis. By the time the author became a professor, he was asked to be a member of committees involving the industry and research and development institutions. He was fortunate to have the opportunity to hear about the issue of coolant void reactivity of the demonstration reactor of heavy water-moderated boiling water light water-cooled pressure tube reactor and passive safety light water reactors. He also had the opportunity to hear about light water reactors at the advisory council for safety review of nuclear power plants. The author was also in charge of the joint use and operation management of the University of Tokyo's fast neutron source reactor "Yayoi", so he had many opportunities to learn about fast reactors through that research. The author also conducted neutron shielding experiments for the development of fast reactors. Since he was working in Tokai Village, he had the opportunity to learn about various research reactors and test reactors of the Japan Atomic Energy Research Institute (JAERI, now JAEA). The power reactors of Japan Atomic Power Company were located a few hundred meters from the research facility of the University of Tokyo where the author was working, and they were adjacent to the site boundary. When the author was an undergraduate at the university, he was taken care of at the Calder Hall reactor, which had just started operating, during his summer internship. Since becoming an associate professor, the author had many opportunities to travel to the USA and Europe for the American Nuclear Society meetings and international conferences, etc. During these trips, the author was fortunate to visit research reactors and power reactors, and learn about various reactors and their research and development. In the 1990s, the author was in charge of exchanging information on fast reactors between Japan and Russia for several years, and the author also visited various reactors at Russian research and development institutions. One of his two mentors was involved in the design and construction of the JPDR, the first light water reactor test reactor in Japan, and was an expert in light water reactors. He was well aware of the design and evolution of light water reactors, and his words were very helpful not only in the class but also after becoming a faculty member.

In the own design research, the author started with a medical reactor for boron neutron capture therapy. When young, the author studied nuclear fusion reactors and nuclear fusion-fission hybrid reactors. The graduate students supervised were excellent and were able to create their own analysis codes for design research. We even did calculations for target implosion for the conceptual design study of inertial fusion reactors. When he became a professor, the author noticed the advantage of making light water reactors supercritical, and studied the concept of supercritical pressure light water reactors (SCWR) with associate professors, assistants, and graduate students for 25 years. It was necessary to consider while referring to the design and analysis of light water reactors, etc. which was useful for understanding light water reactors. Not only studying, but also designing and analyzing supercritical pressure light water reactors while referring to light water reactors, the author advanced his understanding of the design and analysis methods of power reactors.

The supercritical pressure light water reactor was selected as the Generation IV reactor in 2002. The author was invited to advice the European SCWR study. He also received competitive funds in Japan and conducted design research on supercritical pressure light water-cooled fast reactors. The experiments were commissioned to universities and research and development institutions. Since it does not cost to conduct conceptual design research by ourselves, the thermal neutron reactor of the supercritical pressure light water reactor was researched as a unique theme of the laboratory. We did not need to receive external funds including scientific research funds for the study. The students in the laboratory were excellent, and they researched as the theme of their graduation thesis, master's thesis, and doctoral thesis. In Japan it was not necessary to financially support them. It might be said that it was a good time. In addition, scientific research funds were necessary for the operation of the research group, so they were acquired on a theme separate from design study.

There exist other books and reports that summarize the history of power reactors. However, the unique feature of this book is that it describes facts from a bird's-eye view, showing the basis of the literature, from light water reactors to fusion reactors, and not just describing the facts, but also expressing considerations from the perspective of reactor physics, heat transfer fluid engineering, safety analysis, and safety design as a reactor expert. It would be appreciated if the reader does not take the consideration results for granted, but reads the materials shown as references and thinks about them if they have different opinions.

Another feature of this book is that it explains the whole concept design and analysis of power generation reactors, taking the supercritical pressure light water reactor as an example. It also discusses the knowledge base for use and human resource development related to use and research and development in university education. It also touches on the social aspects of the development and use of power reactors, such as investment risks. The social aspects of nuclear power utilization are discussed in the sister book of Springer, "Nuclear Power and Society: Economy, Safety, Environment and Law".

For the future, it is necessary to understand the whole picture and lessons of past development. Experience will be lost if not passed on. The author hopes that both books will be used as textbooks and supplementary reading materials in university classes. When using them as textbooks, please use the references as well, because the description of the books is concise.

In order to advance the use of nuclear power, it is necessary to face the history and lessons of the past, recognize the problems, avoid repeating the same mistakes, and think about solutions. The author expresses his gratitude to those who have helped and cooperated in various ways, and hopes that this book will be useful for the future use and research and development of nuclear power generation in the world.

The manufacturer's authorised representative in the EU is Springer Nature Customer Service Centre GmbH, Europaplatz 3, 69115 Heidelberg, Germany. If you have any concerns regarding our products, please contact ProductSafety@springernature.com

Printed and bound by CPI Group (UK) Ltd, Croydon, CR0 4YY
01/10/2026
02225928-0001